Apiary (Apiculture)

OrangeBooks Publication

Smriti Nagar, Bhilai, Chhattisgarh - 490020

Website:**www.orangebooks.in**

First Edition, 2022

ISBN: 978-93-92878-18-3

APIARY
(APICULTURE)

Dr. YOGENDRA KUMAR PAYASI,
PINKI DUDHWAL
Dr. ANURADHA SHUKLA,
Prof. J.P. SHUKLA

OrangeBooks Publication
www.orangebooks.in

Preface

Human beings have always been looking towards living fauna to utilize them as food, shelter, recreation, and other agricultural purposes since time immemorial. The pharmaceuticals obtained from animals and their products have nice response in cure of various diseases without affecting the longevity of life. Various life saving drugs like cardiac glucosides, insulin etc. have been obtained from animals. Though some animals have been found useful to mankind while others cause greatloss to the human's economy.Though each and every animal species has its own significance from biodiversity view point, however, some of them like birds, reptiles, pisces and arthropods in the current are of technology ere gaining too much importance. Few animals exhibit standard status in small scale industries benefiting to rural people of our nation.

In new education policy commenced on 19 July 2020, directions and suggestions have been given to improve rural poverty by boosting the scientific approaches from lab too land. In this continuation certificate courses, diploma courses and degree courses have been stated to bring it more effective for rural people.

Keeping this in view, an attempt have been made to enhance the production of honey bee products without any complex technology like honey, bee venom, royal jelly, bee waxe etc by culturing honey bees.The scientific rearing of honey bees is called apiculture or Apiary.

We hope that the students and beekeepers would be able to get salient information's on bee keeping through this book entitled "Apiary". Any suggestion for the improvement of the book would be the thankfully accepted.

Authors (JPS and YKP) are grateful to their colleagues, students and friends for their valuable cooperation in the preparation of this book. They are highly obliged to VC, S.P.M. TripathiJi, elderly Shri B.D. SharmaJi and Professor Poonam Sharma, Head Department of Zoology for their encouragement and mental boost.

One of the authors (JPS) dedicates the entire credit to his heavenly resting teacher Professor Kamleshwar Pandey and Parents, Late Shri R.C. Shukla (Father) and Saroja Shukla (Mother). Also one of the author (Anuradha Shukla) gives credit of composing this book to her Late grandfather Shri T.N. Shukla whose blesses inspired her composing the few part of this book. Again one of the authors (YKP) dedicates this book to heaven resting father Late Prof. S.K. Payasi. Pinki Dudhwal heartly thanks to her father Sri P. R. Dudhwal and teachers for blessings and moral support.

Authors also thanks to Mr. Amit Kumar Shiv and Vineet Kumar Mourya for correction typing and to publisher and his production department for publishing this book.

YK Payasi
Pinki Dudhwal
Anuradha Shukla
J. P. Shukla

Index

UNIT - 1

Basics Of Apiculture

Synopsis-(Introduction; History of bee-keeping; Biology of Honey bees; Hive maintenance; Seasonality; Swarming; Egg laying rate; Honey bee classification; Species of Honey bees; Social organization of Honey bee colony; Bee development (Life history); Principles of Bee keeping and Queen rearing).

Introduction:

Apiculture or apiology is the science of rearing the honey bee for the commercial production of honey and other products like wax, pollen, bee venom and royal jelly. It is also called beekeeping. Beekeepers are known as apiarist and place where bees are maintained is called as Apiary, or Bee yard.

Bee keeping is called apiculture, as the honey bee belongs to the genus *Apis*. Honey bees suck nectar from flowers and collect it in their hives naturally. In addition to honey, beehives are a source of wax that is used in various medicinal and commercial preparations.

Many attempts have been made during 1655-1809 in England, Germany and France for organized apiculture which resulted into significant progress in this field. Pokrovic of USSR, first and foremost used boxes for bee rearing in 1878. Thereafter, nearly fifty types of bee keeping boxes were constructed in Europe and America. In India, modern bee keeping was started in 1894 by John Doglas working in Postal department. He also authored a book on Apiculture in entitled "Handbook of Bee keeping."

Honey, has been under use in human civilization since prehistoric period as cited in our religious literatures like Purans, Vedas Ramayan, Mahabharat and Charak Sangheeta. The highly evolved social organization of honey bees had been established before the existance of human race. the bee keeping in US, Canada, New Zeeland and Australia has achieved out-standing success.

History of Bee Keeping-

As cited in Ramayana, King Sugreev had maintained a honey forest at Kiskindha mountain where honey was produced by techniques of bee keeping. Some famous authorities in bee keeping are:

- Johann Dzierzou is known as the father of Apiculture (Apiology).
- Moses Quinby is known as the Father of Commercial Beekeeping.
- Langstroth is known as the father of American Apicellure.
- Rev. Francis Newton of Tamilnadu is known as the Father of Indian Apiculture. Newtont was a faculty member in St. Joseph college Trichurapalli (Tamilnadu).

About 40-30 million years ago during Eocene they appear, to have developed social behavior and structurally and virtually are identical with modern bees.

Honeybees probably originated in Tropical Africa and spread from South Africa to Northern Europe and West into India and China. They were brought to the Americans with the first colonists and are now distributed worldwide. The first bees appeared in the fossil record in deposits dating about 40 million years ago.

Biology of Honeybees

Honey bees are one of the few insects that have a social structure, a caste, which consists of a single reproductive queen (only egg layer in the colony), numerous drones (males) depending upon time of year, and a small number to about 60,000 worker bees or non-reproductive (sterile) female bees. Honey bees undergo complete metamorphosis (holometamorphosis) and develop through four life cycle stages normally, egg, larva, pupa, and adult. Several thousand worker bees cooperate in nest building, food collection, and brood rearing. Each honey bee in the worker caste has an age-related task to perform, which begins inside the hive (bee house) and eventually moves to foraging outside the hive. The immature forms of the bee are called the brood, and they are fed and cared for by the worker bee caste.

Hive Maintenance

Like any good home, a honey bee hive needs to be maintained in order to produce a healthy and productive brood. Construction and maintenance of honey bee hives is conducted by the worker bee caste. Young adult honey bees convert excess food energy in their bodies for wax production rather than fat production. Wax glands are located on the underside of the abdominal segments, and the bees use their legs to scrape off the produced wax and use it in honey comb production. Worker bees collect propolis from tree buds and use it like caulking to seal cracks and leaks around the hive. During hot summer days, the colony temperature must be reduced because the wax may melt and the bees suffer, so they collect water and spread it on the interior of the nest and fan their wings, thereby, causing evaporative cooling. As adult bee grow in age, they perform the following tasks in this order: clean cells (house bees), circulate air with their wings, feed larvae, practice flying, receive pollen and nectar from foragers, guard hive entrance and eventually, they will move out of the hive and forage.

Seasonality

As the level of pollen, nectar and water sources change in the environment, the activity and maintenance of the hive changes accordingly in response The changing conditions in the hive are enhanced in cold winter environment, and they are not as pronounced as in tropical and sub-tropical environments.

1). Spring

During spring when the days are longer and sources of pollen and nectar appear and increase, they stimulate egg laying by the queen and brood rearing. As such, the colony population increases, the worker force increases, and the number of foragers increase. Surplus collection of pollen and nectar are then stored to maintain brood rearing.

2). Summer

Day length in the summer is the longest and bees can then forage for extended periods, again collecting additional stores of pollen and nectar. However, in arid regime, a reduced supply of nectar and pollen is available in the natural environment. The colony reaches peak population in the early summer.

3). Fall

Hive populations diminish in the fall in response to a reduction in the amount of nectar and pollen collected and reduced brood rearing. Additionally, the proportion of old bees in the colony decreases and is dependent on the age, health and fecundity of the queen.

4). Winter

Whenever it's cold, the bees will cluster around the eggs, larvae, and pupae and keep them warm through heat generated by the bees. Under subtropical, tropical, and mild winter conditions, egg laying and brood rearing are usually not affected.

Swarming

The mass flight of a part of honey bees from a large colony to start a new colony is called swarming. The mass of honey bees making a swarm is called a swarming consisting of queen, drones and worker in thousands of number without a comb. The first swarm is led by the old queen but the second swarm is led by the 7-8 days old virgin queen which is followed by the drones and is called nuptial or marriage flight. One of the drone begins populating with the queen in open sky and fertilizes the queen and dies during the course of population.

During early winter, the queen increases drone production by laying unfertilized eggs. This is in preparation of colony swarming. Workers are also likely to prepare to rear a new queen. A few early stage larvae are fed a special food called royal jelly and their cells are enlarged to accommodate the larger queen. Eventually, the original queen will leave the hive and a large number of worker bees will accompany her in search of a new hive location. After flying around in the air for several minutes, they will typically cluster on the twig or branch of a tree or similar object, but they won't stay there long. They depart the next day in search of a good place to start making a new hive. Swarming generally occurs in the Central, Southern, and Western States from March to June, although it can occur at almost any time from April to October.

The remaining bees continue to work as normal with the exception of the care of the new queen larvae. When a new queen' emerges from her cell, she searches for rival queens and destroys the cells. If a rival has emerged from its cell, the two queens will fight until one has eliminated the other. There will be only one queen. When the new queen is about a week old, she will mate with one or more drones outside the colony in the air. Within 3 or 4 days the mated queen begins egg laying.

Egg laying rate-

One queen lays about 1500-2000 eggs in a day depending upon the season and other ecological factors. In a entire span of life of 2 - 5 years, a single queen lays about 15,00,000 (Fifteen lacs) eggs. It is shall not be out of place to mention that when queen in a colony loses her egg laying capacity, another worker of the same colony starts feeding on the queen's i.e. Royal Jelly and developes into new queen and is provided with all the facilities of real queen. Sometimes, when 2-3 queens are developed in a colony, only single takes the position of the real queen and the others, come out with some workers to establish new colonies.

Classification of Honey bee

Despite thousands of years of exploration by humans, honeybees are often mistaken and misidentified. Bees, wasps, and ants are all part of a group of insects called Hymenopetra meaning "membrane-winged" from the Greek word hymen (membrane) and pteron (wing). There are currently more than 100,000 different Hymenoptera species (Jones, Sweeny-Lynch, 1980). This large group of diverse insects is broken down into two groups:

*Symphyta (sawflies and horntails) which have a broad junction between thorax and abdomen.

*Apocrita (ants, bees, and wasps) which have a narrow junction between the thorax and abdomen.

Genus -Apis

Suba family - Apinae

Within a Apocrita, honey bees are classified into the family Apidae," which includes all bees and wasps. Further, they are classified as a member of the genus Apis, distinguished by the production and storage of honey and the construction of "nests" from wax. Bees are included in Sub-family Apinae.

The family Apidae is large and includes many different winged wasps and bees that are often mistaken for honeybees. Despite the colors, and often structural similarities, honeybees have a unique appearance and lifestyle, which sets them apart from their relatives.

- Honeybees have relatively plump and fuzzy bodies. Their hind legs are flattened.
* Bumblebees are much larger, plumper and fuzzier than honeybees.

* Yellow jackets, have smooth, bright yellow and black bodies with a well-defined waist and thin legs. The bee genus Apis occurs naturally in Europe, Asia, and Africa.

Honey bee classification

Phylum- Arthropoda

Class - Insecta or Hexepoden

Order- Hymenoptera

Family – Apidae

Sub family- Apinae

Genus - Apis

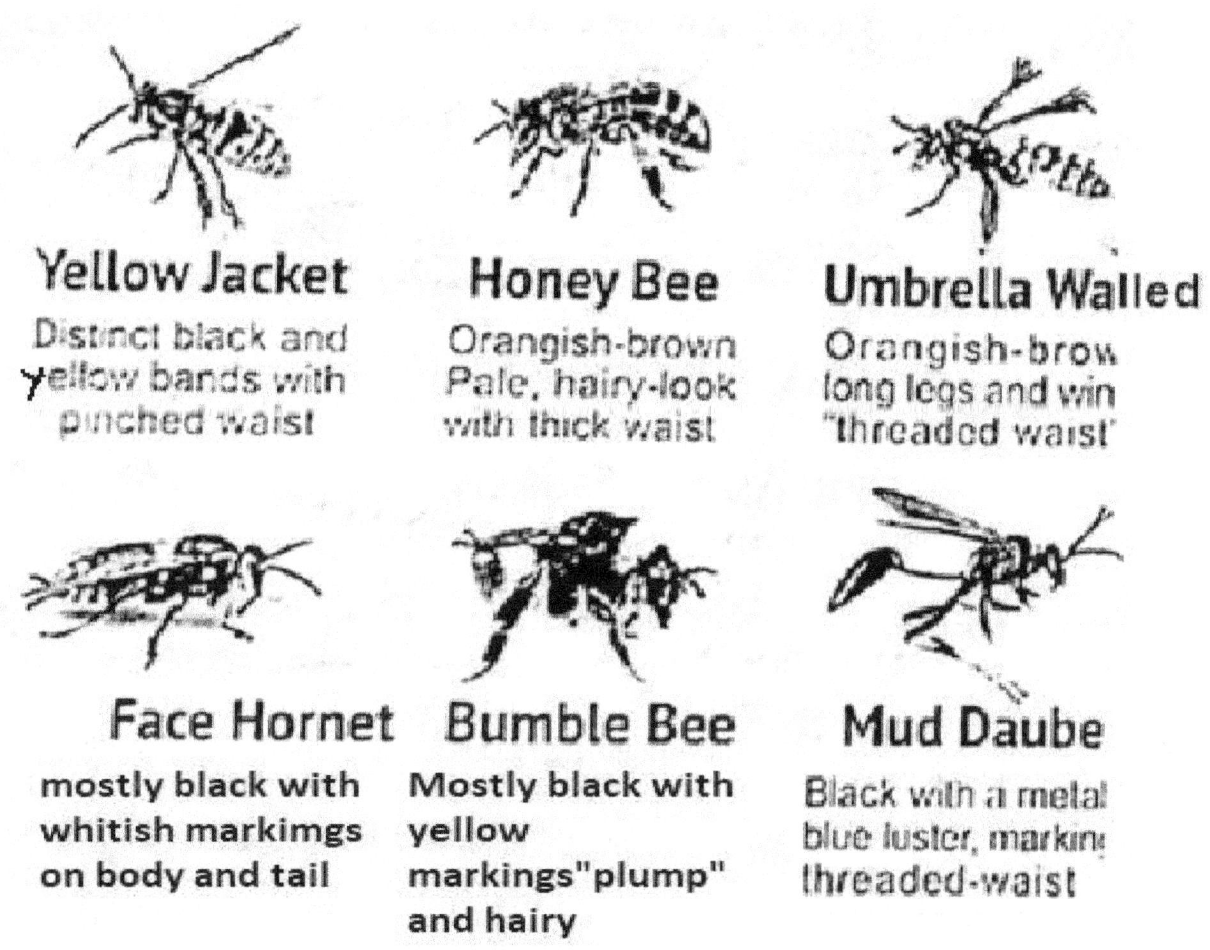

Fig.1.1 Different Kinds of Honey bee

Species of Honeybees

About 20,000 species of bees exist in nature, however, only eight species of honey bee are recognized, with a total of 44 subspecies, although historically 7 to 11 species are recognized – *Apis andreniformis* (the black dwarf honey bee); *Apis cerama* (the eastern honey bee); *Apis dorsata* (the giant honey bee); *Apis florea* (the red dwarf honey bee); *Apis koschevnikovi* (Koschevnikov's honey bee); *Apis laborosa* (the Himalayan giant honey bee); *Apis mellifera* (the western honey bee); and *Apis nigrocincta* (the Philippine honey bee). However, *Apis indica*, earlier called as *A. cerana* is a popular indian bee used in apiculture.

These are all united by traits of honey production, wax comb production, and living in a colony with a queen. They also demonstrate unique behaviours and characteristics that should be noted by keen beekeepers.

Some important species of honey bees for production purposes are as follows -

1. Apis dorsata (The rock bee)

This honey-bee having reddish colour is found all over India upto the height of 12,000 m in the hilly regions. It is the largest sized (20 mm) Indian bees and also known as Sarang or Giant bee or Bombara.

The bees build hives in tall trees, rocks, caves and ceilings of the deserted buildings. Forewing is the longest among all the honey bees which is about 12.0-14.5mm They are migratory in habit and move to the hilly areas in swarm during June to July and return to the plains in winter. The comb is single having hexagonal cells. It is composed of way.

They can build large-sized hives which are about 1.5m to 2.Om from side to side and above 1.0m from top to bottom. Though a single hive produces up to 36-40 kg of honey during a year but the bee-keepers have rarely been successful in rearing due to the wild and ferocious nature of the worker bees. The colony strength is about 90000 bees. The comb is largest among all the honey bees (1.8-5.0 m across).

Cubital Index of Apis dorsata is 6.1-9.8mm. Cubital index is the ratio of the two of the wing vein segments of honey bees. The pattern of the veins of forewing is specific for each breed of the bees. "

It is to be noted that Tomenta may be narrow, or median or broad in most of the bee species. It varies from 3-6 in all species except in Apis mellifera, it may be -3-5. Tomanta refers to the width and length of the patches of over hairs that occur on the abdominal tergites (dorsal plate) of bees.

2. Apis cerana or Apis indica (The Indian hive bee)
They are found all over India both in the plains and hilly areas. The bees make hives in the side tree holes, thick bushes and on the mud walls and they mostly prefer to live in the dark protected places. They also construct comb in rock cavities, and hollows tree trunk. Comb is multiple and comb cells are hexagonal. Annual honey yield is 2.0-5.0 kg.

The workers in the plain region are yellowish white and at high altitudes are dark in colour. They are gentle in nature, so bee-keepers use widely Apis indica in apiculture. A single comb produces 2.0 kg honey in' the plains and about 5.0 kg in the hilly regions. Cubital index of A indica is 3.1-5.0 mm.

It is smaller than rock bee but larger than little bee or dwarf honey bee (A florea). It's yellowish in colour .

APIS INDICA   *APIS MELLIFERA*

APIS FLOREA

APIS DORSATA

4 Salient species of Honey bee in relation to length

3. Apis florea (The little bee)

It is also called as dwarf honey bee. Its forewing is the shortest among all species of honey bees. They are mainly found in the plains of India. The bees are smaller than the other Indian bees. The queen is golden brown.in colour and the drones are black with greyish hair.The comb is exposed and hangs from twigs. They make hives in tree holes, on the branches of trees and rock crevices. They frequently change their places and prefer to remain wild. The hive is considerably small and produces small amount of honey (about 0.5 kg annualy). Colony strength ranges from 40000-50000 bees. Comb diameter is about 20 cm. Cubital index is 2.8-3.7 mm.

4. Apis mellifera (The European or Italian bee)

They are found in Africa, Europe, America, and Canada and have also been introduced all over the world including India. It is smaller than rock bee bust larger than little bee. It is dark brown in colour. The comb is umexposed and is constructed in cavities. The colony strength is 80,000 – 90.000 bees.

They are also called European bees as they are native to Europo. They are about 8 - 9 mm in size and build multiple combs in dark cavities, caves and crevices. They are docile and produce about 10 - 30 kg of honey per annum They can also be effectively used for bookeeping.

Cubital under of A mellifera is 1.65 - 2.95 mm

Note: Besides the above described four species of honey bee, one more species also yields too little amount of honey. It is *Trigona iridipennis.* It is commonly called as dammer bee. It is very small of all the honey bees and is stingless. Its colour is black and comb is unexposed. Comb cells are ellipsoidal. Comb is made up of cerunum (a mixture of wax and resin). The length of comb is about 25-30cm, breadth about 10-11 cm and height about l0- 12cm. Anual honey yied is 65-120ml.

Social organization of Honey bee Colony

Honey bees are social insects, which means that they live together in large, well-organized family groups. Social insects are highly evolved insects that engage in a variety of complex tasks not practiced by the multitude of solitary insects. Communication, complex nest construction, environmental control, defense, and division of the labor are just some of the behaviours that honey bees have developed to exist successfully in social colonies. These fascinating behaviours make social insects in general, and honey bees in particular, among the most fascinating creatures on earth. It shall not be out of place to mention that plants on earth planet depend upon bees. It is because about 75 percent of agriculture is pollinated by honey bees. According to Albert Einstein, if bees disappeared from earth, humanity can not survive more than four years. If we save the bees, then we save our planet. In simple words if there is no bee, then there is no food.

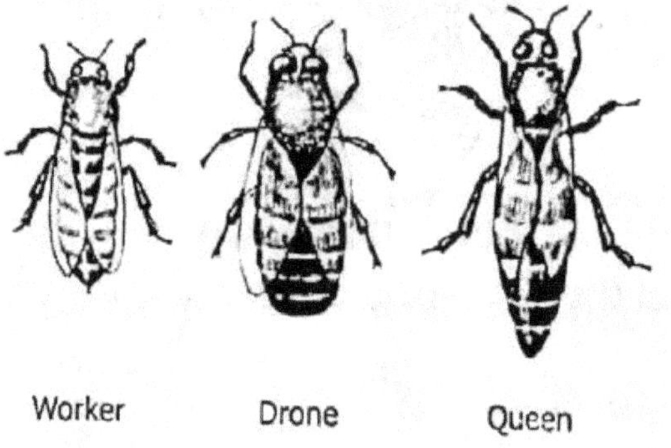

Fig 1-2. Three castes of Apis indica

A honey bee colony typically consists of three kinds of adult bees: workers, drones, and a queen. Several thousand worker bees cooperate in nest building, food collection, and brood rearing. Each member has a definite task to perform, related to its adult age. But surviving and reproducing take the combined efforts of the entire colony. Individual bees (workers, drones, and queens) cannot survive without the support of the colony members in the form of caste (Worker, Drones and Queen).

In addition to thousands of worker adults, a colony normally has a single queen and several hundred drones during late spring and summer. The social structure of the colony is maintained by the presence of the queen and workers and depends on an effective system of communication. The distribution of semio-chemical, pheromones among members and communicative **"dances"** are responsible for controlling the activities necessary for colony survival. Labor activities among worker bees depend primarily on the age of the bee but vary with the needs of the colony. Reproduction and colony strength depend on the queen, the quantity of food stores, and the size of the worker force. As the size of the colony increases up to a maximum of about 60,000 workers, so does the efficiency of the colony.

Queen -

Each colony has only one queen, except during and a varying period following swarming preparations or supersedure. Because she is the only sexually developed female, her primary function is reproduction. She produces both fertilized and unfertilized eggs. Queens lay the greatest number of eggs in the spring and early summer. During peak production, queens may lay up to 1,500 eggs per day. They gradually cease laying eggs in early October and produce few or no eggs until early next spring (January). One queen may produce up to 250,000 eggs per year and possibly more than a million in her lifetime.

A queen is easily distinguished from other members of the colony. Her body is normally much longer than either the drone's or worker's, especially during the egg-laying period when her abdomen is greatly elongated. Her wings cover only about two-thirds of the abdomen, whereas the wings of both workers and drones nearly reach the tip of the abdomen when folded. A queen's thorax is slightly larger than that of a worker, and she has neither pollen baskets nor functional wax glands. Her sting is curved and longer than that of the worker, but it has fewer and shorter barbs. The queen can live for several years, sometimes for as long as 5 years but average productive life span is 2 to 3 years(maximum 5 years),

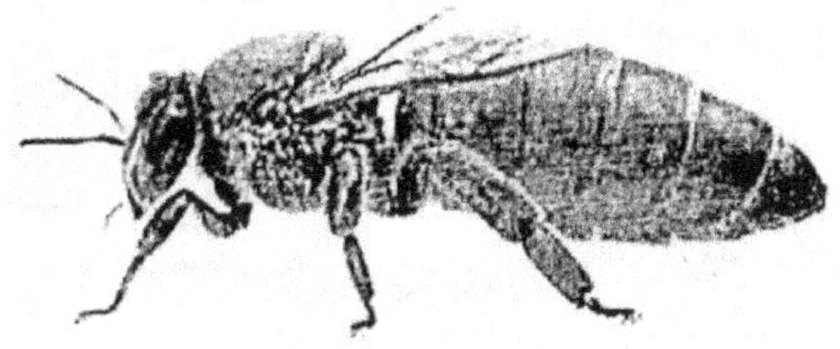

Queen bee

The socond major function of a queen is producing pheromones that serve as a social "glue" unifying and helping to give individual identity to a bee colony. One major pheromone termed as queen substance-is produced by her mandibular glands, but others are also important. The qualities of the colony depend largely on the egg-laying and chemical production capabilities of the queen. Her genetic makeup-along with that of the drones she has mated with,contributes significantly to the quality, size, and temperament of the colony.

About one week after emerging from a queen cell, the queen leaves the hive to mate with several drones in flight. Because she must fly some distance from her colony to mate (nature's way of avoiding inbreeding), she first circles the hive to orient herself to its location. She leaves the hive by herself and is gone approximately 13 minutes. The queen mates, usually in the afternoon, with seven to fifteen drones at an altitude above 20 feet. Drones are able to find and recognize the queen by her chemical odor (pheromone). If bad weather delays the queen's mating flight extends for more than 20 days, she loses the ability to mate and will only be able to lay unfertilized eggs, which results in drones.

After mating the queen returns to the hive and begins laying eggs in about 48 hours. She releases several sperms from the spermatheca each time she lays an egg destined to become either a worker or queen. If her egg is laid in a larger drone-sized cell, she does not release sperm. The queen is constantly attended and fed royal jelly by the colony's worker bees. The number of eggs the queen lays depends on the amount of food she receives and the size of the worker force capable of preparing beeswax cells for her eggs and caring for the larva that will hatch from the eggs in 3 days. When the queen substance secreted by the queen is no longer adequate, the workers prepare to replace (supersede) her. The old queen and her new daughter may both be present in the hive for some time following supersedure.

New queens develop from fertilized eggs or from young worker larvae not more than 3 days old. New queens are raised under three different circumstances: emergency, supersedure, or swarming. When an old queen is accidentally killed, lost or removed, the bees select younger worker larvae to produce emergency queens. These queens are raised in worker cells modified to hang vertically on the comb surface. When an older queen begins to fail (decreased production of queen substance), the colony prepares to raise a new queen. Queens produced as a result of supersedure are usually better than emergency queens since they receive larger quantities of food (royal jelly) during development, Like emergency queen cells, supersedure queen cells typically are raised on the comb surface. In comparison, queen cells produced in preparation for swarming are found along the bottom margins of the frames or in gaps in the beeswax combs within the brood area.

What is Royal Jelly?- The hypopharyngeal glands in nurse (worker) honey bee are helpful in making the royal jelly as well as making the honey comb. After 3 days of development, the workers and drones bees do not feed on royal jelly. However, the queen feeds on this substance throughout its development. Since, royal jelly does have antibacterial and antifungal properties queen bees live 40 times longer than worker bees. It shall not be out of place to mention that royal jelly is the junk, developing bees float in it until they metamorphose.

Drones (male bees) are the largest bees in the colony. They are generally present only during late spring and summer. The drone's head is much larger than that of either the queen or worker, and its compound eyes meet at the top of its head.

Royal Jelly is a complex mixture of various substances in different quantity Nagai & in one 2004). It is a milky gelatinus secretion produced by worker honey bees. It is to feed the queen bee and their young. After the larval stage is complete and the queen bee energes, she is fed royal jelly throughout her life. Royal jelly is used in the nutrition of larvae and adult queen. Royal Jelly is fed to all larvae in the colony, regardless of sex or caste.

Chemical Composition of Royal Jelly:
Water: 50-60%

Protein: 12-18 %

Carbohydrates (monosaccharides): 10-16%

Fatty acids : 3 – 6 %

Minerals, water soluble vitamins : Trace amount

Salient water soluble vitamins in Royal Jelly are - Thiamine (Vit B1), - Riboflavin (Vit B2), Niacin (Vit B3) Pantothenic acid (Vit B5), Pyridoxine (Vit B6) & Biotin (Vit. B7), Inositol (Vit B8) and Folic acid . No fat soluble vitamin is found in Royal Jelly. (MRJPs) are a family of proteins (major royal jelly proteins), secreted by worker honey bees. This family consists of nine proteins, of which MRJP 1 (also called Royal actin), MRJP2, MRJP 3, MR JP4 and MRJP5 are present in the royal jelly secreted by worker bees. MRJP1 is the most abundant and largest in size. The five proteins constitute 83-90 percent of the total protein in the royal jelly.

Drones

Drones are the male bees of the colony. They are larger than worker but smaller than queen. They appear generally during late spring and till summer. The drone's had is much larger than that of the queen or worker. The compound eyes are located at the top of the head. Drones become sexually mature about a week after emerging and die instantly after mating. Their main function is to fertilize the queen.

Drones have no sting pollen baskets, or wax glands. Their main function is to fertilize the virgin queen during her mating flight. Drones become sexually mature about a week after emerging and die instantly upon mating. Although drones perform no useful work for the hive, their presence is believed to be important for normal colony functioning.

While drones normally rely on workers for food, they can feed themselves within the hive after they are 4 days old. Since drones eat three times as much food as workers, and excessive number of drones may place an added stress on the colony's food supply. Drones stay in the hive until they are about 8 days old, after which they begin to take orientation flights. Drones have never been observed taking food from flowers.

When cold weather begins in the fall and pollen/nectar resources become scarce, drones usually are forced out into the cold and left to starve. Queenless colonies, however, allow them to stay in the hive indefinitely. The drones are the male members of the colony and fertilized the queen hence called as King of the honey bee colony. The take about 24 days to develop from the egg to become adult. The sting and the wax glands are absent but the male reproductive system is well developed. They are reared from an unfertilized egg in a large Drone cell. In fact, the drones are fully dependent upon the workers and have been found begging for honey from the worker. The sole duty of the drone is to fertilize the virgin queen. During swarming, the drones follow the queen, copulate and die after copulation.

Worker

Workers are the smallest and constitute the majority of bees occupying the colony. They are sexually rather undeveloped females or sterilized females and under normal hive conditions do not lay eggs. Workers have specialized structures, such as brood food glands, scent glands, wax glands, and pollen baskets, which allow them to perform all the labors of the hive. They clean and polish the cells, feed the broad, care for the queen, remove debris, handle incoming nectar, build beeswax combs, guard the entrance, and air-condition and ventilate the hive during their initial few weeks as adults. Later as field bees they forage for nectar, pollen, water, and propolis.

The life span of the worker during summer is about 6 weeks. Workers reared in the fall may live as long as 6 months, allowing the colony to survive the winter and assisting in the rearing of new generations in the spring before they die.

Although the workers are the smallest of the three castes but they function as the main spring of the complicated machinery like honey bee colony. Like the queen, they are also produced from the fertile eggs laid by the queen and live in a chamber called as 'WORKER CELL'. It takes 21 days in the development from the egg to the adult and the total life span of a worker is about 6 weeks. The workers are atrophid female which sacrifice themselves for the well-being of the colony. The total indoor and outdoor duties of the colony are performed by the workers only. That is why they are provided with some special structures for particular work.

(1) Long proboscis for sucking the nectar.

(2) Strong wings for fanning.

(3) Pollen baskets for the collection of pollen.

(4) Powerful sting to defend the colony against any attack.

(5) Wax gland for wax secretion.

The workers which are engaged in outdoor duties, collect the nectar, pollen, gum and water which are received and stored properly by the house bees. The indoor workers are further sub-grouped for specific duties. Some of them which are very sincere attend the queen while some others look after the nursery called as NURSERY BEE. Some produce wax for the formation of the new hive and are known as BUILDERS. The repairing of the comb is done by the REPAIRERS. The dead body and other impurities are removed from the hive by the CLEANERS. The worker bee which performs fanning in the hive is called FANNER. Many other functions like storage of honey and ripening are also done by the workers. The guard bee always watches at the gateway. It is accepted that upto half of their life period, workers perform indoor duties and further on become engaged in outdoor duties. When a colony becomes queenless, the ovaries of several workers develop and workers begin to lay unfertilized eggs. Development of the workers ovary is believed to be inhibited by the presence of brood and the queen and her chemicals. The presence of laying workers in a colony usually means that the colony has been queenless for one or more weeks. However, laying workers also may be found in normal "queenright" colonies during the swarming season and when the colony is headed by a poor queen. Colonies with laying workers are recognized easily: there may be anywhere from five to fifteen eggs per cell and small-bodied drones are reared in worker-sized cells. In addition, laying workers scatter their eggs more randomly over the brood combs, and eggs can be found on the sides of the cell of t instead of at the base, where they are placed by a queen. Some of these eggs do not hatch, and many of the drone larvae that do hatch do not survive to maturity in the smaller cells.

Bee development - [Life History]
All three types of adult honey bees pass through three developmental stages before emerging as adults: egg, larva, and pupa. The three stages are collectively labeled as brood. While the developmental stages are similar, they do differ in duration. Unfertilized eggs become drones, while fertilized eggs become cither workers or queens. Nutrition plays an important part in caste development of female bees; larvae destined to become workers receive less royal jelly and more a mixture of honey and pollen compared to the copious amounts of royal jelly that the queen larva receives.

Eggs
Honey bee eggs are normally laid one per cell by the queen. Each egg is attached to the cell bottom and looks like a tiny grain of rice. When first laid, the egg stands straight up on end. However, during the 3-day development period, the egg begins to bend over. On the third day, the egg hatches into a tiny grub and the larval stage begins. The eggs are pinkish in colour, elongated with cylindrical body.

The Lifecycle Of The Honey Bee

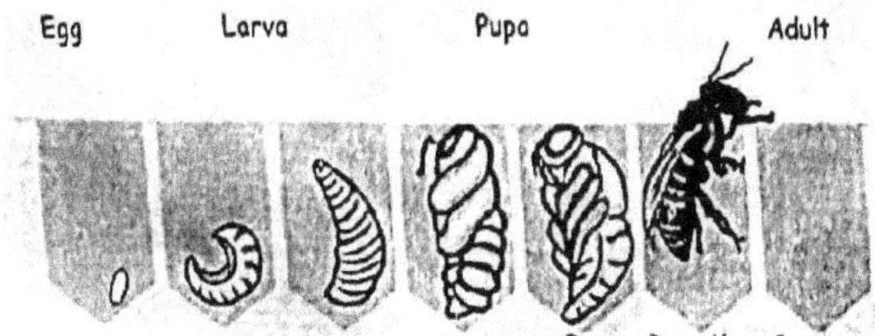

Fig. 1.3 Three different stages of life cycle (from egg to Adult)

Larvae

Larvae emerge out from both the fertilized as well as unfertilized eggs. The larvae from unfertilized eggs form the drones whereas the larvae from the fertilized eggs develop into workers. The larvae of the worker feed on royal jelly and becomes the queen of the colony and rest become sterile females (workers). Royal jelly constitutes a special diet consisting of digested honey and pollen mixed with glandular secretion in to the mouth of the worker.

Healthy larvae are pearly white in color with a glistening appearance. They are curved in a "C" shape on the bottom of the cell. Worker, queen, and drone cells are capped after larvae are approximately 5, 6, and 6 1/2 days old, respectively. During the larval stage, they are fed by adult worker bees while still inside their beeswax cells. The period just after the cell is capped is called the prepupal stage. During this stage the larva is still grub-like in appearance but stretches itself out lengthwise in the cell and spins a thin silken cocoon. Larvae remain pearly white, plump, and glistening during the prepupal stage.

Pupae

Within the individual cells capped with a beeswax cover provided by adult worker bees, the prepupae begin to change from their larval form to adult bees. Healthy pupae remain white and glistening during the initial stages of development, even though their bodies begin to take on adult forms. Compound eyes are the first feature begin to take on: color, changing from white to brownish-purple. Soon after this, the rest of the body begins to take on the color of an adult bee. New workers, queens, and drones emerge approximately after 12, 7 and 14 days, respectively, after their cells are capped.

Principles of Apiculture "(Beekeeping)"

The rearing of honey bees for the commercial production of honey, wax, venom, and royal jelly is called apiculture, or beekeeping .The success of bee keeping relies on the following principles:

1. The aim and objective of Apiculture should not based the production of honey alone; But it should be aimed on the production of the following products namely:

* Honey

*. Wax ~

* Bee bread

* Royal jelly

* Bee venom

* Propolis

*Queen for sales

* Crop yield through pollination

The profit is much more on crop yield through pollination than that of honey alone. The following are the salient points for apiary.

* The apiary should be arranged in an area rich in bee flora. Bee flora are plants on which bees are foraging. Plant more bee flora, of different special.

*The area should be free from insecticides.

* The apiary should be protected from enemies, wind, smoke etc.

* Bees should reared in modern hives.

* The hives should be arranged in a line or in a semicircle.

* The hives should be arranged in groups.

*They are erected in areas exposed to sun light.

* Wood materials should be used for hives.

* Swarming should be checked.

* Selection of disease resistant bee stock.

*The life cycle and behaviour of bees should be known to the beekeeper.

* *Apis Indica* and *Apis mellifera* are proper for bee keeping purpose

* As apiculture is an Agroindustry, the ecobiology of the bees should be leant.

*The queen should be out standing in her quality because of the following reasons. She is the mother of the colony. Drones and workers are sons and daughter respectively. She contributes about 50% of the genome of the colony. The life span of worker bees is about 6 weeks. Hundreds of then die every day. This loss is compensated by queen laying about 1500-2000 eggs per day. Pheromones released from queen integrates and coordinates all the members of the colony. If the queen becomes old then it should be substituted by new queen for the colony. Artificial feeding using corn syrup or sugar syrup should be made on the lean swarm. Hive must be inspected weekly. Extraction of honey from brood comb should be avoided. During processing, honey should not be heated above 35°C. Honey packing should be done in air tight containers.

Queen Rearing

Beekeeping can be advanced when queen rearing is promoted. Queen rearing is a process of mass production of queens from young worker larvae by a queenless colony. Its main object is to enhance bee Keeping for its various products. Queen rearing is also termed as queen raising which may be natural or artificial.

In natural queen rearing, the queenless colony produces three types of queens namely Swarm queen; =and emergency queen.

The artificial queen rearing is by two methods namely - 1. Non-grafting method and 2. Grafting method.

In non-grafting method, the worker larva is reared into queen in the same cell where it is hatched. The cell is enlarged into a queen cell and the larva is fed with royal jelly by nurse bees Non-grating, method includes the following - 1. Hopkins method; 2- Miller method and 3 - Alley method.

In grafting method, one-day worker larva is. shifted or transferred to a queen cell by the bee keeper. The nurse bees feed the larva with royal jelly and the larva is reared to a queen by the nurse bees. It is called Doolittle method. It is done in the following steps:

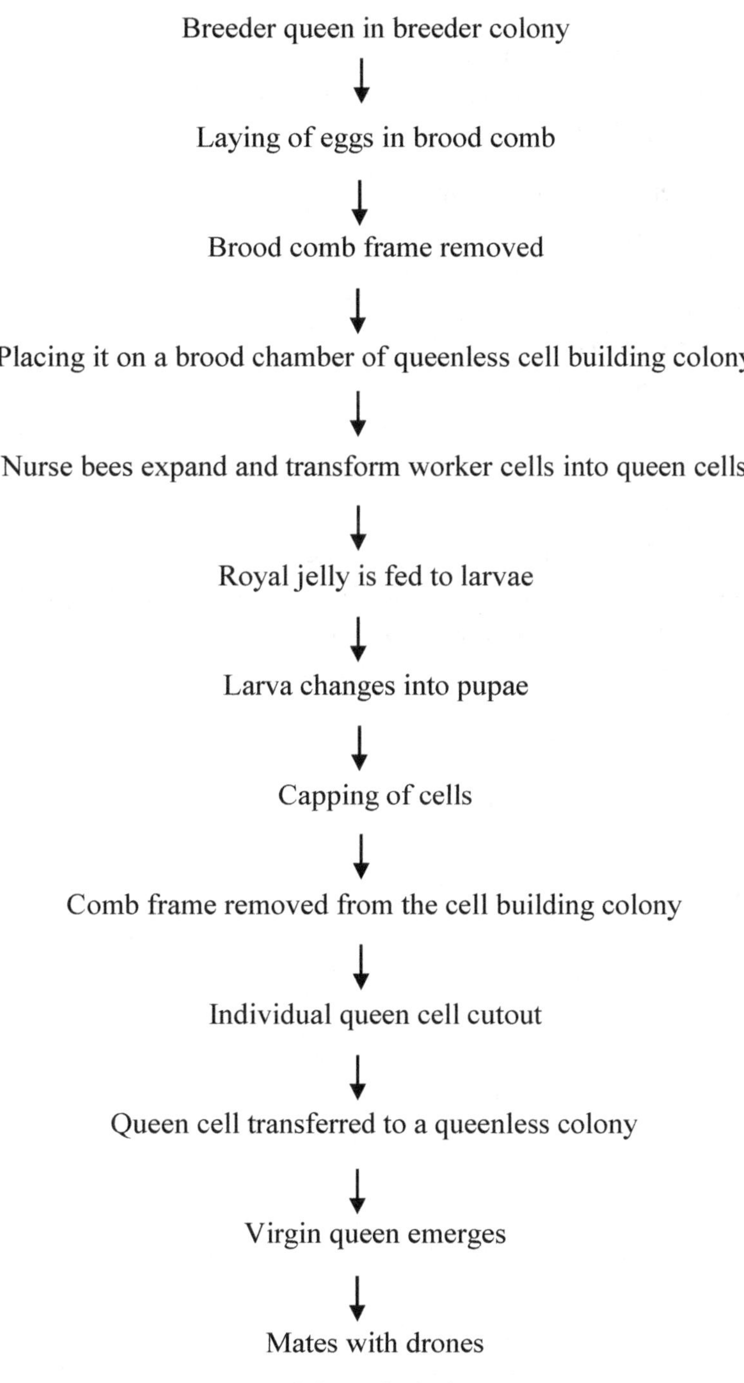

Fig.1.4- Doolitle method of queen rearing

Need for Rearing of queen

She contributes 50 percent of the genome of the colony. The life span of worker bee is only about six weeks. Hundreds of them die every day. This loss is compensated by queen laying about 1500-2000 eggs per day.

* Pheromones released from queen integrates and coordinates all the members of the colony
* If the queen becomes old, it should be substituted by new queen for the colony.
* Artificial feeding using corn syrup or sugar syrup should be made on the lean season.
* Hive must be inspected weekly.
* Extraction of honey from brood comb should be avoided.
* During processing, honey shall not be heated above 35 degree centigrade.
* Honey packing must be in air tight containers.

The salient points for of queen are the following
* To requeen a queenless colony.
* To replace and infected and diseased queen .
* To replace and old queen.
* To produce a superior queen for new colony.
* Desirable queen trait can be achived.
* For selling to bee keeprs.

Important Questions-

1. Define bee keeping. Give its history and importance.
2. Give an account of different species of Honey bee and classify the Indian bee.
3. Give an account of principles of bee keeping and queen rearing.
4. Describe various events in the life history of *Apis indica.*
5. Write short notes on the following:
   A. Egg laying rate of *Apis indica*
   B. Hive maintenance
   C. Biology of Honey bee
   D. Queen rearing
   E.HoneybeeColony

UNIT - 2

Bee-Hive And Methods Of Bee Keeping

Synopsis:- (Introduction; Types of bee hives (traditional or primitive hive, modern hive and movable hives); Flora of Apiculture; Selection of bees for Apiculture; Desirable traits of bees for choice in Apiculture; Methods of bee keeping and extraction of Honey).

Introduction

Bee-hive is the place a honey bee colony called home. The bee family may occupy a man-made wooden box or an empty hollow log. The ability of bees to find new places to build their hives is truly fascinating. If conditions change or the colony outgrows their current location, they will search out a new hive to frame home.

In fact, the house of honey bees is termed as hive or comb. It consists of hexagonal cells made up of wax secreted by the worker's abdomen. These hives are hanging vertically from rock, building or branches of trees. Each hive has thousands of hexagonal thin walled fragile cells arranged in two opposite rows on a common base. The resins and gums secreted from the plants are used for the repairing of the hives. The young stages are generally occupying the lower and central cells in the hive which are the 'BROOD CELLS'. In *A. dorsata* broad cells are similar in shape and size, but in rest of the honey bee species, brood cells are of 3 kinds namely, Worker cell, Drone cell and Queen cell respectively. Queen cell is used for single time but the rest are used many times. The lower part of the comb are used for brood rearing and the upper part for the storage of honey.

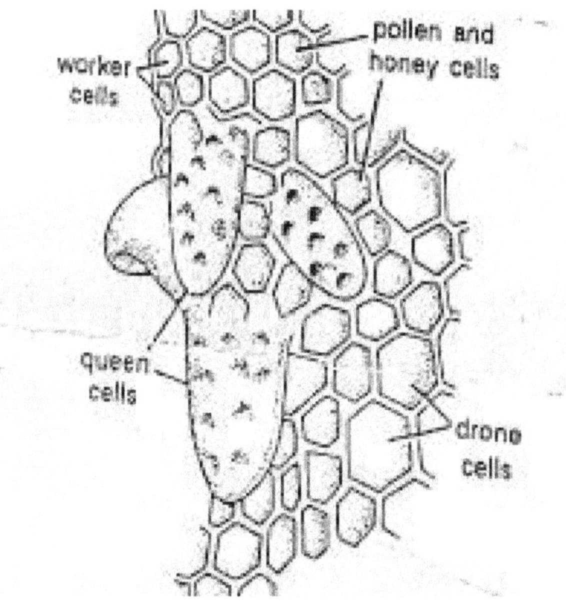

Fig. 21 Hive of Apis indica

Types of Bee-hive

A bee hive is an enclosed structure in which the honey bees are reared. Bee hives are grouped into two:

(1) Traditional hives or Primitive hives

(2) Modern hives or movable hives.

(1). **Traditional hives or Primitive hives:-** These are the primitive bee keeping methods and are fixed comb hives usually. The hives used for bee keeping in non-scientific era (ancient time) are called as traditional or primitive hives. Traditional bee hives are simply provided with an enclosure for the bee colony. Because no internal structures were provided for the bees, the bees created their own honeycomb within the hives. The comb is often cross-attached and cannot be moved without destroying it. This is sometimes called a fixed-frame hive to differentiate it from the modern movable-frame hives. Harvesting generally destroys the hives, though there may be some adaptations using extra top baskets which could be removed when the bees filled them with honey. These may be gradually suplemented with box hives of varying dimensions, with or without frames, and finally replaced by newer modern equipment.

Honey from traditional hives is typically extracted by pressing, crushing the wax honeycomb to squeeze out the honey. Due to this harvesting, traditional bee hives typically having more wax but less honey.

The hives used for beekeeping in non-scientific era (ancient times) are called as primitive or traditional hives.

Following are the types of traditional bee hive:

(1). Mud and Clay hive

(2) Wall hive

(3) Pot hive

(4) Skeps hive

(5) Log hives

(6) Bamboo hives

(7) Bee gums hive

(8) Barrel hives

(9) Gourd hives

1). Mud and Clay hives

Mud hives are extended, long cylinders made with a blend of dung, straw, and unbaked mud while clay hives are built of baked clay, primarily used for maintaining a reduced number of colonies. One end of the hive is smoked to drive the bees out.

2) Wall hives

It is a fixed type man made hives having rectangular cavities in the wall of the homes. It contains a small hole for the entrance of honey bees and a wide hole with cover for the collection of comb. In this type, the honey bees are attached with the combs on the inside of the hive. During the harvesting, honey bees are destroyed and the combs are cut out. Collection of the honey is done by the crashing of the combs.

3) Pot hives

The pots applied as hives are called Pot hives. It is practiced in Coorg, Mysore, Malabar, Godawari, Kashmir etc. It can be transferred from one place to another, hence it may be regarded as movable hives. The pots are placed in forests after smearing with wax and sweet scented leaves in the inner surface. Pot hives may be installed on the ground or on pegs in trees. When the bees had finally settled, the pots may be transported to any desired place.

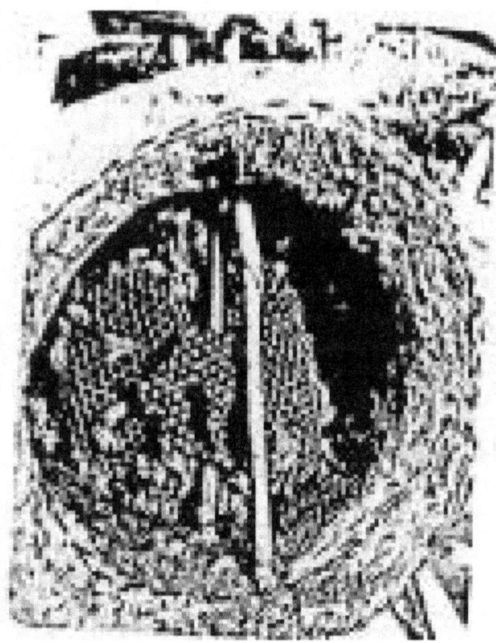

Fig-2.2 Mud & clay hive

4) Skep hive

It is made of grasses or straw; hence also called grass hive. It is used by African tribal people, people of east Asia and Europe. A skep is a traditional round hive made of straw or dried grass. Multiple strands of straw are bundled together to form a thick rope and the rope is then coiled and bound together to make the skep. The skep is generally empty so it provides protection for the bees but little else. There is one opening, usually at the bottom, for the bees to enter. Skeb may be in the form of basket or a cylinder. It has a single entrance at the bottom.

Bees living in skeps may often be killed or driven out so that honey could be collected. Often the entire skep-comb and all-parts are pressed to remove the honey and new skeps are prepared for the following year. It can be used once only.

5) Log hives

Hollow wooden log made hive is called tog hive which may be of movable nature. Plant species, Palmyra Padus is generally used to construct this type of hive. Such hive is used in Ghana, Guinea and Tanzania.

6) Bamboo hives

It is a primitive and movable type hive. Bamboo materials used in making basket are used to construct bamboo hive, it may be of different shape and size.

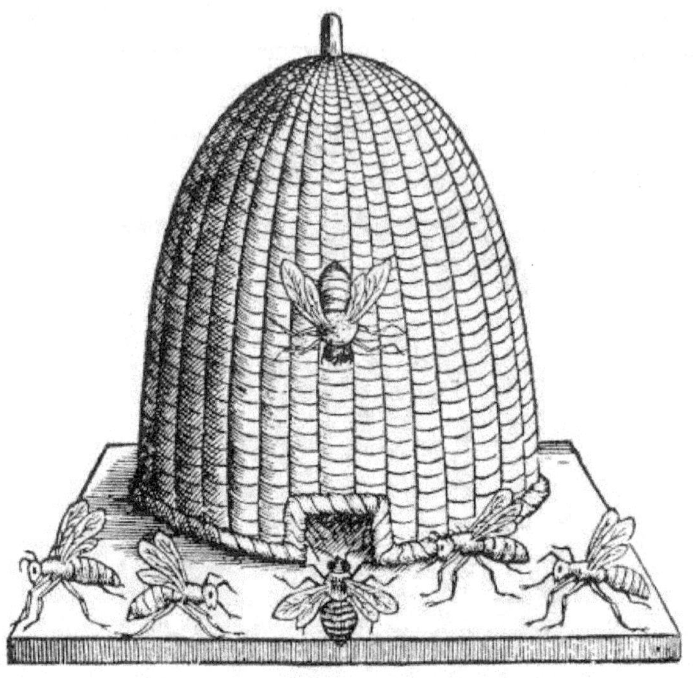

Fig.2.3 Skep Hive (Grass Hive)

7). Bee gums hive

This is a structure in which a hollowed log, mainly from red gum tree, is used for developing the colony. The opening end from the topmost side of the log is usually covered by using wood so that it can serve the purpose of a roof. The hives are destroyed for collecting the honey when it is time of harvesting.

8) Barrel hives

The hive made of wooden barrel is known as barreld hive. It is a primitive type and movable hive. It is not of use in India. It is used in West Africa.

9) Gourd hives

The hives made of dried bottle gourd are called gourd hives. It is a primitive and movable hive. Such hive provides natural hollow for bees. Gourd hive may vary in shape. It may be pot shaped, spherical or long neck pot shaped. It may be hanged or resting on a wooden peg. At the period of harvesting, the entire hive is crushed. Gourd hive is used by tribal people of Savannah and they consume both honey and brood.

2). Modern hives

The bee hives used in the current days are called modern hives. It is modern bee keeping and are movable comb hives. From the movable hives, the comb may be removed and honey may be extracted without damaging the comb. Hence the comb may be reused.

These hives have been modified and improved. From fixed-frame hives, they have been converted into movable structures for better utilization of resource and enhanced flexibility of harvesting.

Such hives can be vertical or horizontal. There are four main types of modern hives in common use worldwide -

1) The Langstroth's hive

2) The Top-bar hive

3) The Warre hive

4) Newton hive

1) The Langstroth's hive

The Langstroth's hive is the most common style in use today and a favorite for new beekeepers. It is included in modern bee keeping method. The design was patented by Rev. Lorenzo Langstroth known as Father of American Apiculture in the mid-19th century (1851) and features removable frames that the bees build comb in. It is suitable for big colonies. It is made up of wood. Langstroth is known as the Father of American Apiculture Langstroth hives consist of boxes that stack on top of each other.

Parts of Langstroth hive :- Langstroth, a modern hive consists of the following parts:

1. Stand

2- Bottom board

3 - Brood chamber

4 - Queen excluder

5- Super chamber

6- Crown board

7- Roof.

1- Stand: It consists of four wooden legs and plateform.

2- Bottom board: It is a wooden plank of about 56 x 41x2.5 cm size placed on the top of the wooden stand. In the centre, it is having air space. It is extended by 10-12 cm in front of the hive body as a landing plateform for bees. It is the base of the hive-It is available as a solid bottom or screened bottom.

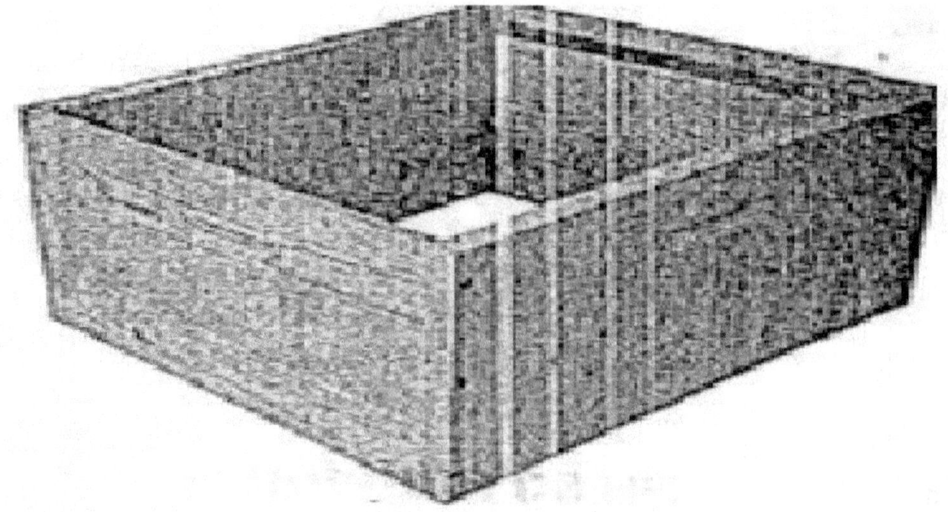

Fig.2.5- Brood Chamber

Fig.2.6- Queen Excluder

Fig.2.7- Super Chamber

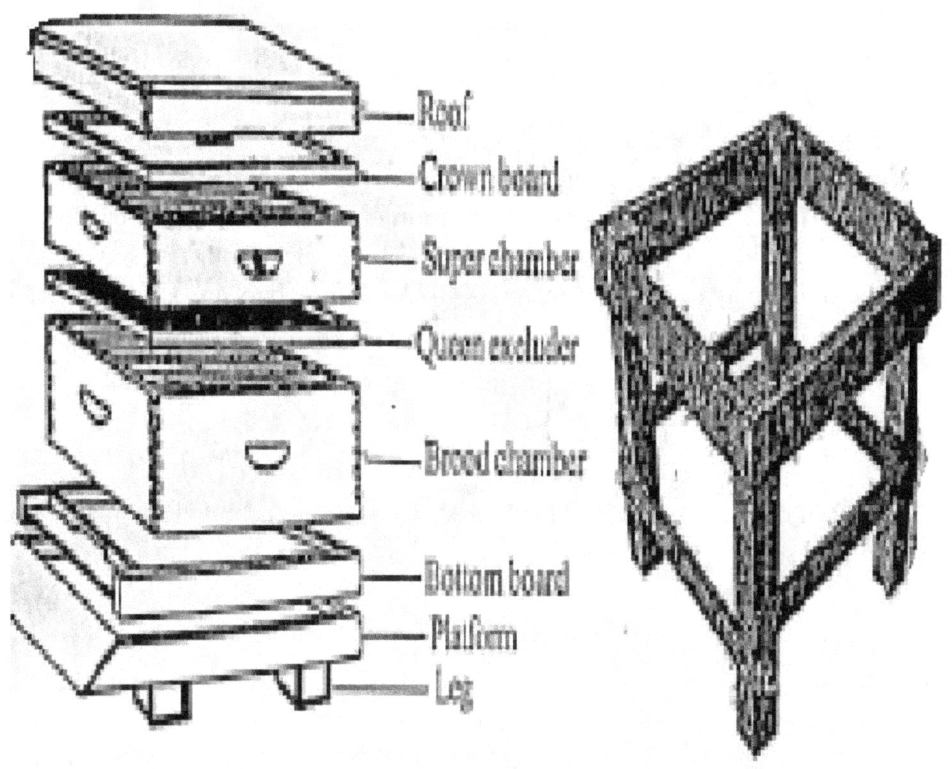

Fig.2.8A- Langstroth's hive

Fig.2.8B- Stand of Langstroth's hive

3). Brood Chamber (deep super or brood box):
It is an open wooden containing brood frames and is placed on the bottom board. It is rectangulate for more of thick wooden plank. The box contains comb frames. The size of brood chamber is 50.5 40.5 x 23.5cm. An entrance is present between the bottom board out the brood chamber. Queen excluder is placed above the brood chamber or prevents the entry of queen into the super chamber. The brood box contains larger frames than the shallow super. Here, the queen lays eggs for the next generation of bees. In this maternity ward, nurse bees care for the young.

4). Queen Excluder
Allows only worker bees to pass through, keeping the queen and drones away from the honey. This is an optional piece of equipment that prevents the queen from laying eggs in the honey collection supers. Not every beekeeper uses an excluder. Queen excluder is kept above the brood chamber. It prevents the entry of queen into the super chamber. It may be a perforated zinc sheet mounted on a wooden frame. The pore size in sheet may be upto 3.5 mm.

5). Super chamber
It is an open wooden box containing honey comb frames. It is placed above the queen excluder. Its common size is 27-28 x 24-25 x 15 cm. Usually super chamber is filled with 10 comb frames. It is used to store the pollen and honey.

6.) Crown board

It is placed above the super chamber and is provided with a central ventilation hole covered with wire gauze.

7.) Roof

It is placed above the crown board and is flat. It is covered over with a metallic thin sheet to make it impervious to rain water. It provides weather protection of the hive. Comb frames are thin rectangular wooden structures and placed in a vertical position in the brood chamber and the super chamber. Each frame consists of a top bar, two side bars and a bottom bar nailed together. Both the ends of the top bar may protrude out so that the frame may rest on the chamber. Sizes of three bars are as follows .

Topbar - 48.0x2.5x2.3cm

Side bar - 22.0x 0.96 x 2.3 cm

Bottom bar - 44.0x0.96x2.3 cm

Normal range of distance left in between two adjacent frames in a bee hive is about 10 mm for Italian bee and about 8-9mm for Indian bees.

2. Top bar hive:

It is a simple rectangular wooden box having horizontal wooden bars on the top. The honey bee fix combs on the bars. The combs hang downwards into the box from the bars. Such combs may not be reused but the bars can be reused. Honey is harvested by removing the bar, cutting the combs and crashing it. The first top bar was designed by Thomas Wildman in 1770. He used extra hives with 7 bars. The top bar hive is the oldest hive design in the world. A horizontal top bar hive features wooden bars that are laid along the top of the long box. One-piece bars are used instead of the 4-sided wooden frames of the Langstroth design. The honey bees build comb down from the top bars. No foundation is required, but the hive should be elevated off the ground with some sort of stand.

There are several advantages to a top bar hive. In addition to not needing foundation sheets, there are no wooden frames to assemble. Perhaps the biggest draw of the top bar hive: no heavy lifting. Unlike the Langstroth hive that requires moving several heavy hive boxes, management of a top bar hive is much easier on the bee keepers back.

Advantages of Top Bar hive: Following are the salient advantages of Top Bar Hive.

- Simple
- Economical
- May be kept as hobby.
- When required, combs may be removed and the honey may be extracted.
- Requires least management

Disadvantages: Top bar hive: Following are the disadvantage of top bar hive
- Combs can not be reused
- Swarming may not be controlled
- Disease control is difficult.
- Requeening is not possible.

Top bar beekeeping does have a few challenges, however for example, a centrifugal honey extractor can not be used to remove honey from the natural comb, so the comb and honey will both need to be removed from the bar. This results in the honey bees having to make new comb each year. In general, top bar hives also require more frequent inspections to prevent overcrowding/ swarming.

Fig.2.9.top-bar hive

This type of hive can produce honey, but it is a favorite for beekeepers wanting hives for pollination alone.

iii). The Warre hive :

The War (war-RAY) hive, created by Emile warre in the mid-20th century, is another top bar design. Instead of being a long horizontal top bar hive, the warer hive is referred to as a vertical top bar hive. Identically sized stacked boxes have no frames or foundation sheets. Bees build honeycomb down from top bars placed within each box.

Fig. 2.10 warre hive

Beekeepers using the waarre style often" bottom Super" their hive instead of putting empty boxes on top to give the colony more overhead roomm, empty boxes are placed at the bottom of the stack. They feel this arrangement on a better mimics bee life in the wild.

4. Newton hive

Newton hive is a wooden box modern and artificial hive used for rearing honey bees. It was designed by Rev. Fr. V. Newton in 1919, of Tamilnadu (India), known as Father of Indian Apiculture. The following are the salient features of Newton hive.

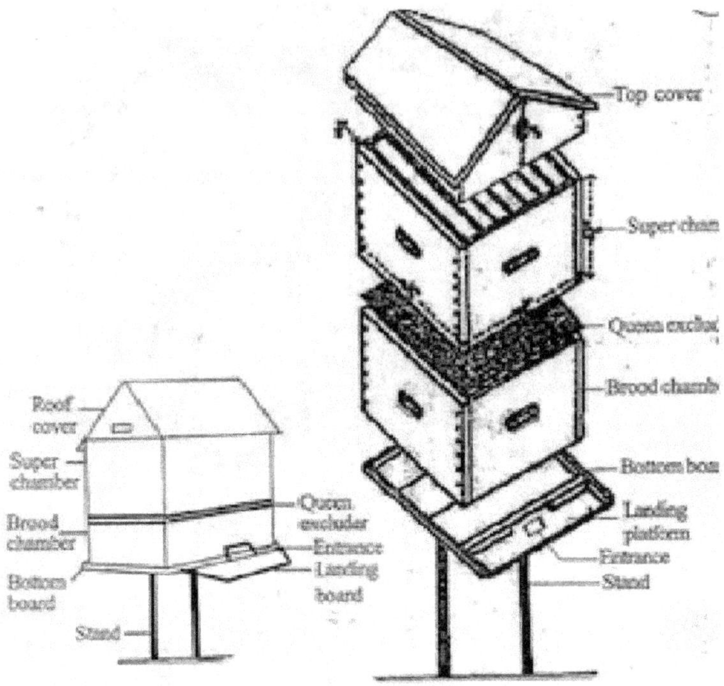

Fig. 2.11 :Newton hive.

- It is a movable frame hive
- It is commonly used by Indian bee keepers.
- It is Useful to rear small colony.
- It is made up of wood.

The Newton hive consists of six main components, namely;

1. Stand

2. Floor board

3. Brood chamber and Brood Frames

4. Queen excluder

5. Super chamber and Super Frames

6. Crown Head

7. Top cover (Roof)

1. Stand- The stand keeps the hive above the ground. It is of two types, namely a pole stand and a leg stand. The pole stand is a single wooden pole of 10cm diameter. The lower end is buried deep in the ground leaving about 25-30cm above the ground. The leg stand has 4 legs. The leg stand is placed on the ground. The pole is fixed deep in the ground. The pole is soaked in wood preservative or a mixture of engine oil and paint thinner to prevent ants and termites. The top of the stand contains a platform and the hive is placed on the platform.

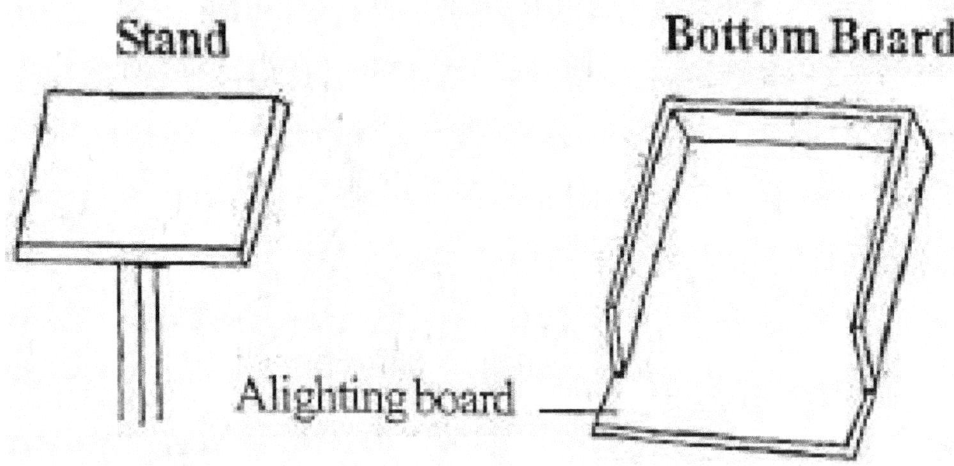

Fig.2.12: Stand   Fig. 2.13: Bottom Board

2. Floor Board- Floor board is a wooden plank placed on the platform. It measures 36 cm x 24 cm x 2.3 cm in size. It has site has an entrance slit of 8.89 x 0.97 cm. The floor board is extended by 10cm in front of the hive body. It provides a landing plat-form for bees.

3. Brood Chamber and Brood Frames-Brood chamber is an open wooden box having brood comb frames. Brood chamber is placed over the floor board.It is a box without top and bottom. It measures about 9.75 x 8.25 x 6,75". The four sides of the chamber are joined by special joints. Between the brood chamber and floor board there is an entrance of 8.89 cm x 0.97 above the floor board. It contains 7 brood frames.

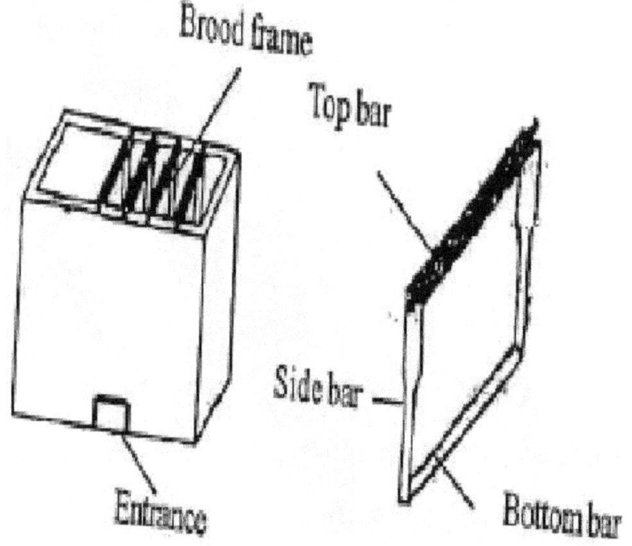

Fig.:2.14: Brood chamber   Fig. 2.15: Brood frame.

A brood frame has a top bar, two side bars and a bottom bar. It appears like a slate. It measures 20.9 x 14.6 x 15.24cm in size and 2.22cm broad. The brood frames contain comb foundation sheet. They are hung inside the brood chamber. It is used for laying eggs and to keep the brood.

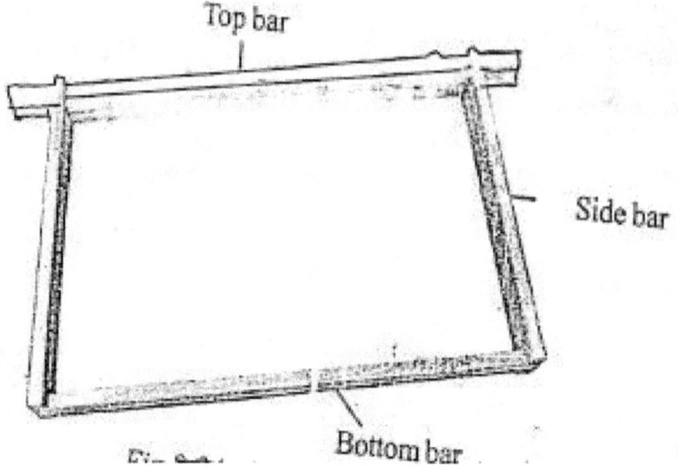

Fig...2.16: A Comb Frame

4. Queen Excluder-Queen excluder is a board applied to keep the queen inside the brood chamber and prevents it to enter the super chamber. It is a perforated zinc sheet mounted on a wooden frame. The size of the pore is about 3.5mm. Queen excluder is kept between brood chamber and super chamber. It can allow the workers to pass through, but prevents the queen entry. It is useful to confine the queen to brood chamber only. It prevents the queen from laying eggs in honeycombs

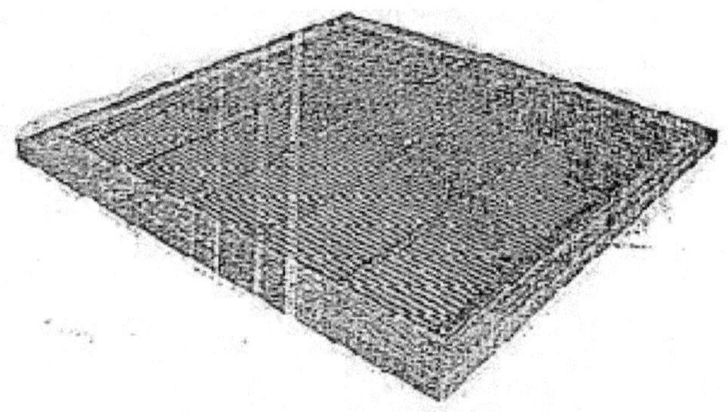

Fig. : Queen excluder.

It is used to prevent the escape of queen from the hive. It also prevents the entry of wasps into the brood chamber.

5. Super Chambers and Super Frame: Super chamber is an open wooden box containing honey comb frames called super chamber because it is placed over the brood chamber. It is simply called super. The super chamber is placed over the queen excluder.

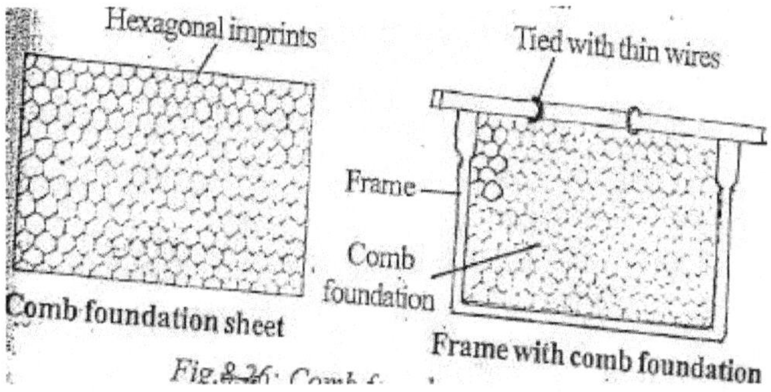

Hexagonal imprints
Tied with thin wires
Frame
Comb foundation
Comb foundation sheet
Frame with comb foundation
Fig.8-26· Comb c

Fig. 2.16: Comb foundation sheet.

It is a box without top and bottom It measures about 9.75 ⅹ 8.25 ⅹ 6.75". It contains super frames.There are 7 super frames. The super frames contain foundation comb sheet. Comb foundation sheet is a thin sheet of bee wax embossed with a pattern of hexagons. It is used to store honey and pollen. Some Newton hive contains two super chambers.

6. Cover (Roof): Top cover is placed above the super chamber. The top cover has two sloping planks. The upper surf ace of the cover is covered with zinc or tin sheets. There are two holes covered with wire mesh on the cover for ventilation.

Advantages of Newton's hive

The hive is fixed tightly to the stand. It facilitates in easy harvesting, and the bee keepers have no fear of damaging the comb. The comb is extracted by means of the centrifugal honey extractor which makes it possible to remove the honey without harming the comb. The combs are reused. During hive operations, very few bees are crushed between frames. The hive is so designed that the queen and brood are confined to the lower chamber. The super chamber is kept on the top. So there is no hindrance for the bees in the brood chamber. Stealing a double or triple storey hive is a tedious task for a any thief.

Bees can easily pass through the various spaces between the frames. Hive boxes can be stacked easily. Drugs can be applied with ease through the openings. Bees are protected from weather and enemies.

Disadvantages of Newton's hive

The following are the disadvantages of Newton's hive:

- A high degree of craftsmanship is required to build the hive.
- A frame hive with two super chambers costs three-four times than the normal hive.
- Only good quality of wood is needed.
- Wood for frame construction must be seasoned for at least a year.
- The need to keep a stock of frames to replace those removed during the honey harvesting creates an additional cost.
- Honey extractors uncapping-knives, trays and other equipments have to be imported
- If frames are unguided, honeybees find it difficult to start the combs correctly to frame.

Flora for Apiculture

Honeybees may collect nectar and pollen from a variety of plants from quite a long distance which are known as bee flora. Nectar is a source of honey, which includes both floral and extra floral nectaries of flower. Pollen is a source of protein. The period when a good number of plants have nectar is called a honey flow period. The season when there is no honey flow is called a dearth period. The honey bees usually forage about 100 m distance from the hive but they can go upto 1.5 km. The bees are most active in foraging within a temperature range of 25c to 27c. The workers make about 6000 trips a day to collect 500 to 1000 mg of pollen. A worker makes about 19000 trips a day for collection of nectar. In addition to pollen and nectar, the bees also collect propils which are a resin like material made by bees from the buds of poplars and cone bearing trees. Salient flora of honey bee are listed ahead in this chapter.

Importance of bee flora : The following are the salient importance of bee flora:

1). Essential to understand various types of bee flora and their blooming phenology in a given area to conserve bee colonies.

2). Pollination is an ecosystem service provided by the bees that is almost always taken for granted.

Choosing flowering plants for a bee. Friendly native garden or Horticulture garden

Native flowers: Honeybees may collect nectar from a wide variety of flowers, but many wild bees are partial to native flowers.

Colorful flowers: Bees are more attracted to yellows and white and the blue-purple range.

Fragrant flowers: Flowers carry fragrance for the very purpose of attracting pollinators.

Long bloomers: Plants with a long flowering season assure the bees of a continuous food supply.

Early season flower: Early spring flowers are a relief to bees that have been on a meager ration during winter.

Mass plantings: Plant large patches of the same plant species in an area, or have a mixed planting with a few selected bee favorites.

List of important Bee flora:

Common name, Botanical name, Family, Flowering period (1-12 months), Source for N-nectar P- pollen) is mentioned below:

1. Stone and pome fruits(*Prunus & Pyrus spp.*), Rosaceae, 2-4 month, N+P

2. Bramble (*Rubus ellipticus.),* Rosaceae, 2-3 month, N+P

3. Barberry(*Berberis lycium*), Berberidaceae, 3-4 month, N+P

4. Honey suckle(*Lonicera angustifolia*), Caprifoliaceae, 3-4 month, NPD

5. Yellow clover(*Medicago denticulata*), Leguminosae, 3-4 month, N+P

6. White clover(*Trifolium repens*), Leguminosae, 3-4 month, N+P

7. Egyptian clover(*Trifolium alexandrinum*), Leguminosae, 4-5 months, N+P

8. Hirad(*Terminalia chebula*), Combretaceae, 4-6 months, N

9. Jamun (*Syzygium cumini*), Myrtaceae, 4-5 months, N+P

10. Eucalyptus(*Eucalyptus sp.),* Myrtaceae, 3-5 months, N+P

11. Bottle brush (*Callistemon lanceolatus*), Myrtaceae, 4 months, N+P

12. False acacia (*Robinia pseudoacacia*), Leguminosae, 4months, N+P

13. Gulmohar (*Jacaranda mimosaefolia*), Bignoniaceae, 4-5 months, N

14. Bird's foot trefoil (*Lotus corniculatus*), Leguminosae, 4-5 months, N.

15. Daru (*Punica granatum*), Punicaceae, 4-5 months, N+P

16. Toon (*Toona ciliata*), Meliaceae, 4-5 months, N+P

17. Sunflower (*Helianthus annuus*), Compositae, 4-7 months, N+P

18. Shisham (*Dalbergia sissoo*), Leguminosae, 4months, N+P

19. Wild rose (*Rosa moschata*), Rosaceae, 4-6 months, N+P

20. Ber (*Zizyphus jujuba*), Rhamanaceae, 5-7 months, N

21. Ohi (*Albizia chinensis*), Mimosaceae, 5-6 months, N.

22. Khair (*Acacia catechu*), Mimosaceae, 5-7 months, N

23. Bhang (*Cannabis sativa*), Cannabaceae, 7-9 months, P

24. Maize (*Zea mays*), Graminae, 8-9 months, P

25. Shain (*Plectranthus rugosus*), Labiatae, 8–10 months, N+P

26. Cruciferous oil seeds (*Brassica spp*), Cruciferae, or Brassicaceae
10-4 months, N+P

27. Wild cherry (*Prunus puddum*), Rosaceae, 10-11 months, N+P

28. Rubber (*Hevea brasiliensis*), Euphorbiaceae, 10-11 months, N

29. Soapnut (*Sapindus spp*), Sapindaceae, 10-12 months, N

Most bees collect just pollen or just nectar on any trip, but a few carry both at the same time. The pollen is stuffed into hair receptacles on their hind legs called corbiculae. Honey bees usually forage on only one kind of flower on any single trip. A single bee can carry about 35% of its body weight of pollen.

Fruit bearing trees -
Most fruit trees provide forage to bees Apple, Plum and Cherry trees are some of the most bee-attracting trees. Fruit trees provide both nectar and pollen to foraging bees in the spring and early summer.

Plants which are visited by bees only for nectar -

- Tamarind (Tamarindus indicus)
- Neem (Azadirachta indica)
- Soapnut tree (Sapindus spp.,)
- Eucalyptus spp.
- Pungam (Pongamia pinnata)
- Moringa tinctoria
- Prosopis spicigera

Plants which supply pollen to the bees

- Sorghum
- Maize
- Roses
- Finger millet
- Bajra
- Castor
- Tobacco

Some beautiful flowers that attract bees -

1. Blackeyed Susan (*Rudbeckia hirta*)- Flowering period is summer blooming flowers.

2. Saffron flower - Saffron are found in Simla , Kulu , Manali regions. Bloom period- October and November.

3. Jamun flower - Jamun fruit flowers (*Naavalpazham*) blossoms in the Himachal region during the month of May and June.

4. Sheesham - *Dalbergia sissoo* commonly known as North Indian Rosewood is an evergreen rosewood tree also known as Sheesham. The flowering seasons are March to May.

5. Mango - In India flowering season start from late December and goes on till March.

6. Ber - Jujube flower (*Elanthap azham*) blossom in the Himalayan and Rajasthan region during the months of September.

7. Aster - It bloom around the time when other flowers are fading. It gives the honeybee a much needed pollination option. They bloom in late summer and stay in bloom.

8. Sunflowers (*Helianthus spp.*) - Sunflowers with their large central discs provide ample opportunity for bees to forage for nectar and pollen.

9. Salvia (*Salvia spp.*) - It provide protein-rich pollen for bees in early spring.

10. Blanket flower (*Gaillardia grandiflora*) - Tough with a profusion of daisy-like flowers can provide sustenance for bees in gardens many other flowers. Bloom period - May to October.

11. Snapdragon - It release a stronger scent than most other flower and this attracts lots and lots of bees.

Honey bees are reared for honey, wax, pollen, bee - venom, pollination, etc. There are about 20,000 known species of honey bees throughout the world. Of these only four species are well known. These are the following:

1. *Apis dorsata* - Rock bee

2. *Apis indica* - Indian bee

3. *Apis mellifera* - Italian (European) bee

4. *Apis florae*- Little bee

Selection of bees for Apiculture

Beekeeping can be taken up with either of the four domesticated honey bee species namely (*Apis indica* , *A. dorsata, A. florea* and *A. mellifera*). However, in cold areas e.g. high hills, *A. Indica* being cold hardy performs better than *A. mellifera*. Moreover, this bee is more frugal and does well even in areas, which are not very rich in bee flora. Farmers who are incapable of making more investment in bee keeping with *A. mellifera* can use *A. indica*, since it needs less investment.

1. *Apis dorsata* (rock bee) - It is also called as giant bee or sarang. They build a single open nest on branches of tall tree, rock and buildings. They are the largest bees, about 20 - 30 mm in size. Each comb may produce 20 - 30 kg of honey. They are highly aggressive and hence cannot be domesticated for beekeeping.

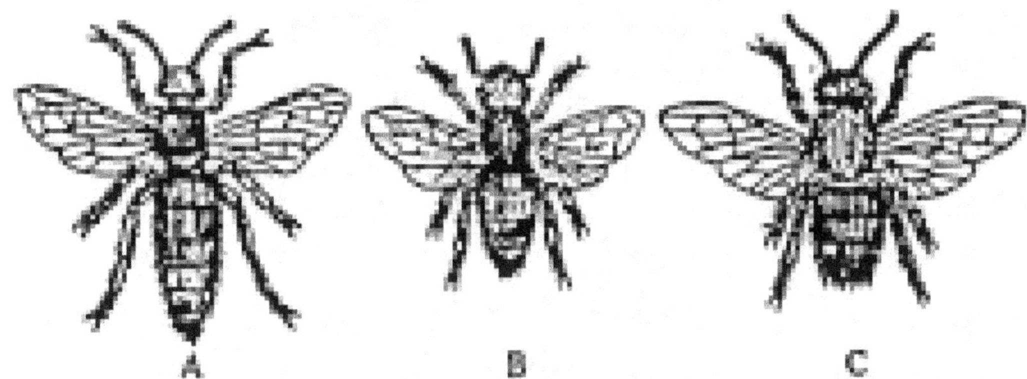

Fig. 2.21. Apis dorsata. A. Queen, B. Worker, C. Drone

2. *Apis florea* (little bee) -. It is also called as dwarf bees. It also constructs single open hive on twigs and branches of shrubs and woody plants close to the ground. They are the smallest, measuring about 5 mm in size and produce only 100 - 250 mg of honey hence they are not suited for beekeeping.

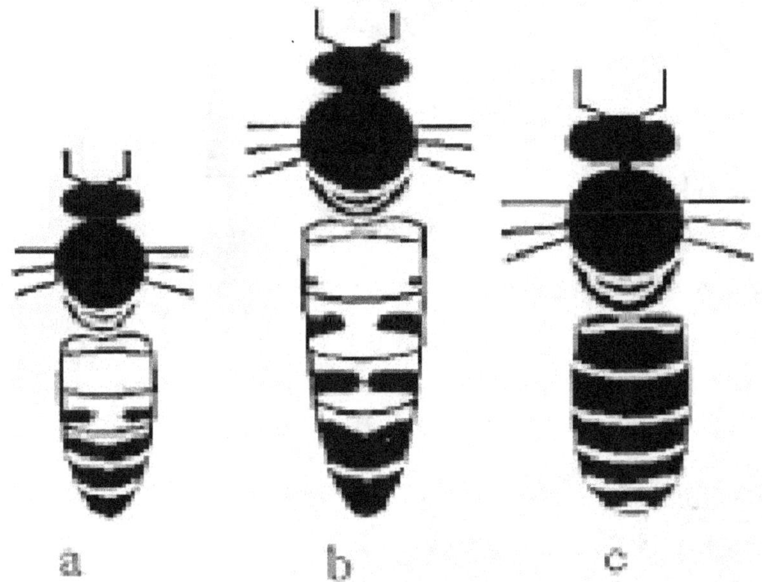

Fig 2.22. Apis florae a. worker, b.queen, c. drone

3. *Apis indica* (Indian bee) - They construct multiple combs in dark cavities or enclosure. They measure about 7 - 8 mm and produce about 3-5 kg of honey in a hive. They are more docile and can be easily maintained by the beekeepers. Hence preferred most in bee keeping.

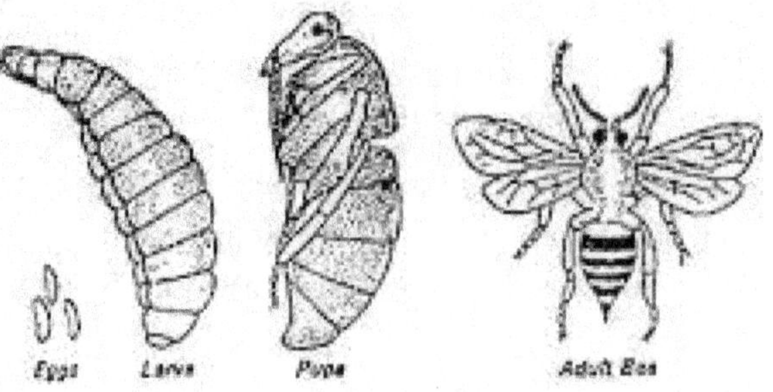

Fig. 2.23.Life history of Apis indica

4. *Apis mellifera* (European boo) - They are also called Italian bees as they are native to Europe. They are about 8 – 9 mm in size and build multiple combs in dark cavities, caves and crevices. They are docile and produce about 10 - 20 kg of honey per annum. They can also be effectively used for beekeeping, with much care.

Fig2.24 Apis mellifera

Desirable Traits for Beekeeping

Apiculture is the rearing of honey bees to obtain honey, wax, bee venom, pollen and other products.

In order to get maximum benefit from apiculture, bees with desirable traits should be kept in the apiary.

-   Bees must build closed nest in unexposed areas.
-   Bees must build multiple nests.
-   Bees must collect large amount of nectar.
-   Honey production must be high.
-   Must be gentle in behavior.
-   Must be calm on combs.
-   Must show low tendency to swarm.
-   Must show low tendency to abscond.

- Must be disease resistant.
- Must use little propolis.
- Must rear little brood during dearth periods.
- They should not be aggressive.
- They must not be irritable.
- The queen must lay more eggs.
- The queen must be superior in genetic make up because 50% of the genome of the colony is contributed by the queen.
- The colony building ability should be high.
- The viability of the egg should be high.
- Brood cycle length should be less for workers.
- Nurse bees should nurture the brood efficiently.
- The workers should be aggressive in foraging.
- The bees should have clustering behaviour during severe cold conditions.

Mode of Hive making and Pollen Collection by Bees -

Hive making is an example of high quality architecture of honey bees. Hive is constructed by workers with wax secreted from the wax glands situated on the ventral surface of their abdomen. The wax is scrubbed by legs and is mixed with the secretion of cephalic gland and then used for making the hive cells with the help of mandible. For scrubbing the wax from the sternal plates of abdomen, a spine is present towards inner side of the tibia of the middle pair of legs. Hive cells are strictly hexagonal. These hexagonal cells are arranged parallel from top to bottom in the hive.

Propolis and Balm- Propolis is actually a resinous substance which is collected by workers from the buds of the flora. It is used for sealing gaps between cells of the hive and repairing the hive. A special type of propolis called balm is for polishing the internal surfaces of the hive.

Pollen Collection

Pollen collection is carried out by workers with the help of their legs which are modified accordingly. Each leg consists of coxa, trochanter, femur, tibia, a tarsus formed of five tarsomeres and pretarsus. A pair of claws are situated on pretarsus.

Tibia of hind leg and basitarsus are basically flat. A groove is present on the outer surface of the tibia which is covered by spines and is called pollen basket. A row of strong spines is present on the terminal end of tibia and is called rake. On the flat inner surface of the basitarsus, are present rows of spines which help in collecting pollens. These spines are called pollen press. There is a protuberence (outgrowth) on the upper end of the basitarsus, called pollen press. Pollen combs act as brushes, to collect pollens from flowers. Pollen grains are separated from the pollen comb by rakes of hind limb and with the aid of pollen press they are transferred to pollen basket. From pollen basket, pollens are transferred to the pollen cells with the help of middle legs. There is a notch on the metatarsus of the middle legs. There is a notch located on the metatarsus of the middle leg which is surrounded by spines around its boundary. A flat protuberance coming out from the terminal end of the tibia, called fibula fits into the notch of metatarsus. Thus a ring like structure is formed which is called toilet organ. This toilet organ is used to remove pollens stuck to antenna and mouth parts (Fig. 2.25)

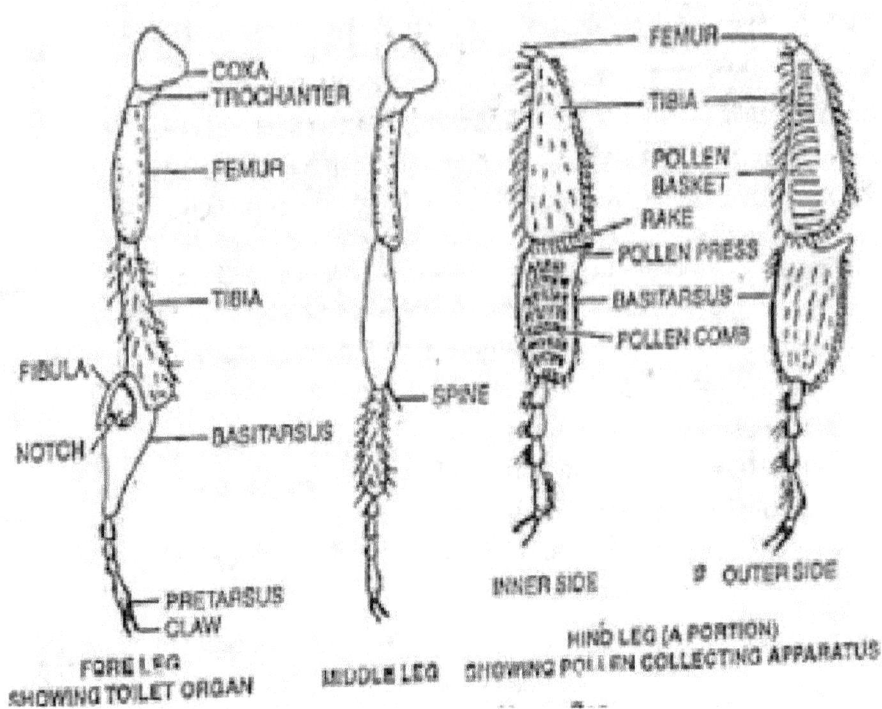

Fig. 2.25. Legs of Honey Bee

Methods of Beekeeping

Beekeeping is a scientific method of keeping *A. cerana* (*indica*) or *A. mellifera* bees for the production of honey and other useful bee products. The main objective is to get more and more quality of honey. There are two methods used by apiculturists. The traditional or **indigenous method** and the modern method.

Indigenous or Traditional Methods of Extraction of Honey

Traditional bee hives are simply provided with an enclosure for the bee colony. Because no internal structures are provided for the bees, the bees create their own honeycomb within the hives, mainly clay hive or mud hive pot. The comb is often cross attached and cannot be moved without destroying it. This is sometimes called a fixed- frame hive to differentiate it from the modern movable hives.

Hive - Two types of hives are used in indigenous method of bee keeping e.g. wall or fixed and movable hive.

Wall or Fixed Hive: It is purely natural type of comb because the bees themselves prepare the hive at any space on the wall or trees. There is an opening on one side through which bees come out of the hive. Various traditional hives have been described earlier in this chapter.

Apiary (Apiculture)

Fig., 2.26 Wall or fixed hive

Movable Hive:

It comprises of hollow wood logs, empty boxes and earthen pots etc. placed in verandas of houses. There exists two holes, one is for entrance and the other for exit of the bees. The swarmed bees usually come to the box on their own accord. Some bee keepers use to take the clusters of the swarms from a tree and keep them in the hive. Movable hive has been described earlier in this chapter"

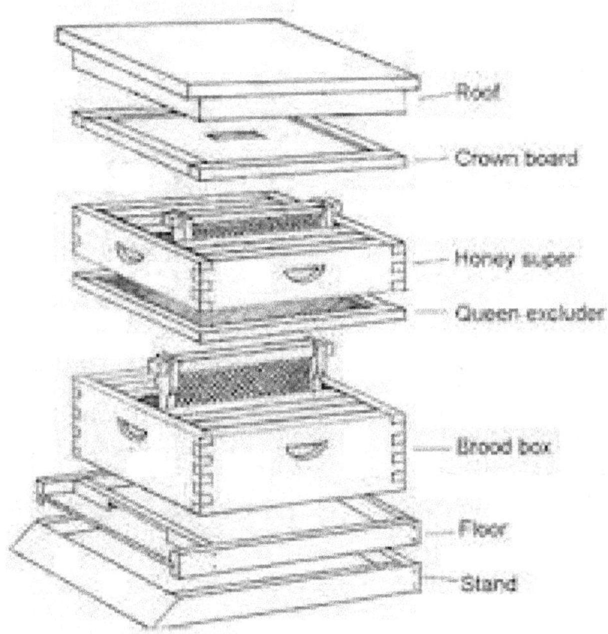

Fig. 2.27 - Other form of movable hive with exposed parts

Extraction of Honey by Honey Extractor: It is used for the extraction of the honey from the comb. It functions on the principle of Centrifugal force. When the combs are centrifuged by honey extractor, the honey is released out without any damage to the comb

For honey extraction, burning fire is brought near the bee hive at the night as a result of which bees are either killed or they escape off.

Further the hive full of honey is being removed, cut into pieces and squeezed to get honey.

Sometimes smoking is done in day period so that the bees may escape from their hives.

Removal of the honey from the honeycomb is called honey extraction. The process of honey extraction needs the following components;

- Honey extractor
- Uncapping knife
- Uncapping tray
- Straining cloth
- Collecting vessel
- Bottles

The honey extraction is done in honey extraction house located away from the hive. The super frames are removed from the hive. The super frames with honey are brought to the honey extraction home. The wax caps are removed from the combs. The removal of cap from the comb is called uncapping. Uncapping is done in the uncapping tray with the help of the uncapping knife.

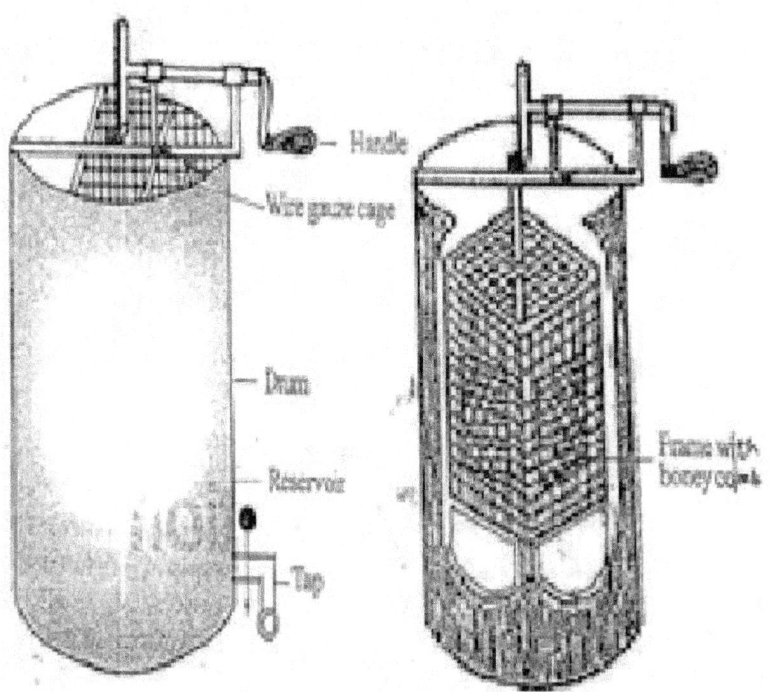

Fig. 2.28.: Honey extractor   *Fig. 2.29: Tangential honey extractor.*

The uncapping knife should be kept warm for easy uncapping, The combs are uncapped on both sides. The uncapped frames are loaded vertically in the wire gauze cage of the honey extractors. A honey extractor is a mechanical device used to separate honey from the combs.

Fig. 2.30: Radial system of extractor

The extractors may be tangential or radial

In the tangential extractor, the combs are arranged facing the cage. But in the radial the combs are arranged radially like the pokes of a wheel.

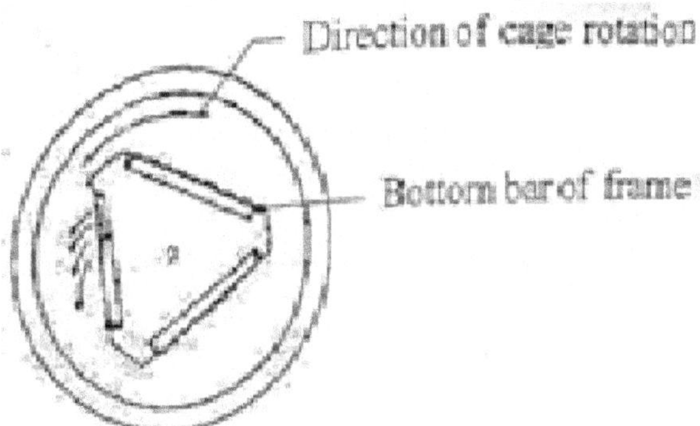

Fig.2.31- Tangential system of extraction

A honey extractor extracts the honey from the honey comb without crushing the comb. Honey extractors work by centrifugal force.

The wire gauze cage is allowed to rotate manually or by a motor.

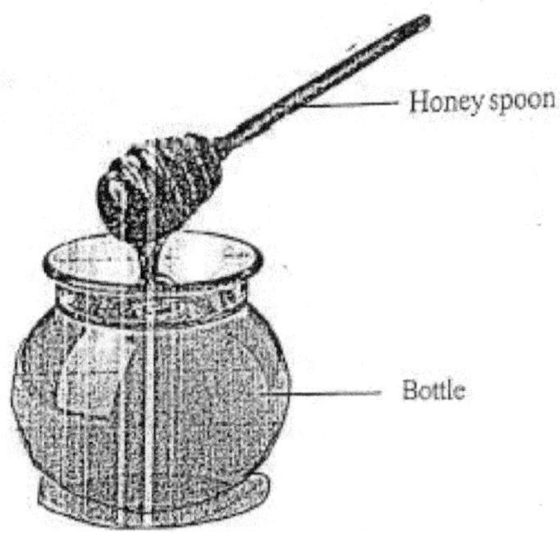

Fig.: 2.32 Bottling of the honey

The rotation must be very gentle at first and the speed of revolution is increased gently.

Rotation process is run for a few minutes.

When the honey combs are rotated, the honey is thrown out into the reservoir of the honey extractor.

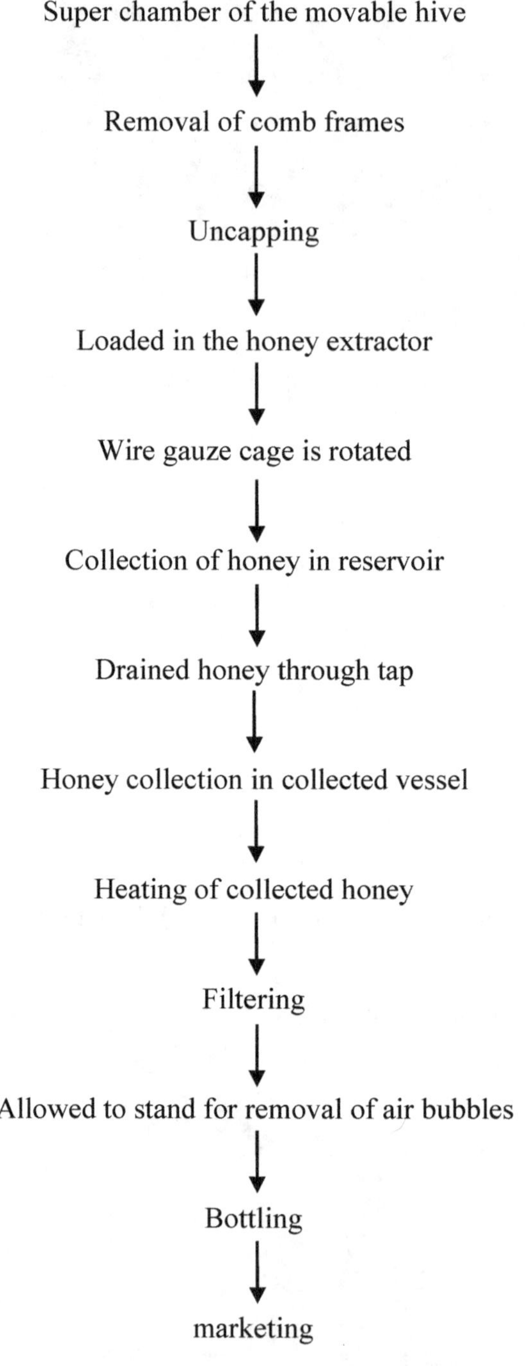

Fig. 2.33: Honey extraction.

Honey gets collected in the reservoir. When the reservoir at the bottom is full, the honey is drained off through in the tap. Honey is collected in a collecting vessels. The honey is heated. Filter the collected honey using a straining cloth. After filtering, the honey is allowed to stand for about 24 hours. This removes the air bubbles. Finally honey is poured into bottles for storage. Lastly honey extraction combs may be reused. Various steps of honey extraction are shown in fig. 2.33.

Questions:

1. Give an account of various types of bee hive.
2. Describe various types of traditional hives.
3. Give an account of various types of modern hives.
4. Describe various flowering flora for Apiculture.
5. Give an account of desirable traits of bee for choice in Apiculture.
6. Give an account of methods of bee keeping.
7. Describe the method for extraction of honey.
8. Write short notes on the following:
   A. Traditional bee hives
   B. Modern bee hives
   C. Flora for bee culture
   D. Extraction of honey
   E. Selection of bees for Apiculture
   F. Method of bee keeping

UNIT - 3
Modern Method Of Apiculture

Synopsis- (Salient aspects of modern bee keeping; Appliances for modern method of bee keeping; Advances of modern method; Advantages of modern bee keeping; Advantages of movable frame hive; Disadvantages of movable frame hives).

Introduction

Modern method of apiculture (bee keeping) is the scientific method of rearing honey bees for honey bee products viz; honey; wax; Royal Jelly; bee bread; propolis; bee venom, Queen; crop yields through pollination etc.

In the modern method of apiculture, the honey bees are reared in movable artificial hives. This was designed and invented by Langstroth in 1951. This invention has turned apiculture into a cottage industry and has provided employment to lakhs of people.

Salient aspects of modern bee keeping:

Modern bee keeping is characterized by the following aspects:

- Aim
- Apiary
- Movable frame hives
- Queen rearing
- Grafting
- Grafting tools
- Artificial queen cups
- Acquiring bees
- Uniting stocks
- Splitting
- Artificial feeding
- Swarm control
- Disease control
- Control of Robbing
- Smoking
- Clothings
- Honey extraction
- Honey processing
- Royal jelly production
- Queen selling

1. Aim of modern beekeeping

Modern beekeeping is aimed not in producing honey alone but in producing following products such as

* Honey
* Wax
* Royal jelly
* Bee bread
* Propolis
* Bee venom
* Queen
* Crop yield through cross pollination

It is to note that the profit is much more on crop yield through cross pollination than that of honey. About 75% agricultural crafts are pollinated by different species by honey bee.

2. Apiary

Apiary is the bee yard where the hives are installed. The success of beekeeping depends on a appropriate site selection. The hives are arranges in series or groups in a line or in a semicircle fashion.

3. Movable frame hives

In modern beekeeping the honey bees are reared in modern hives. The modern hives are of 3 type namely

i. Top bar hives

ii. Langstroth hive

iii. Newton hive

In these hives the comb frames are replaceable. Hence these hives are called movable comb hives.

4. Queen rearing

Queen is known as the mother of colony, the workers and drones of the colony are her daughter and sons respectively. 50% of the genome of the colony is contributed by the queen. She lays about 1500-2000 eggs per day. The life span of worker is about 35 days. Every day hundreds of worker of the hive die. This loss is compensated by the queen laying eggs in enormous number continuously.

The queen coordinates all the members of the colony by secreting queen pheromone (a semio-chemical). The colony dies in absence of queen pheromone. So queen is important for the colony and every year the queen must the replaced. This needs queen rearing.

Queen is reared by 2 methods namely:

1. Non-grafting method

2. Grafting method

In non grafting method, the worker bee larva is allowed to develop into a queen.

In grafting method, the worker larva is transferred to an artificial queen cup and the larva develops into a queen.

The queen rearing produces high quality queens in large numbers.

The queens are used for the following purposes;

i   Requeening.

ii   For producing royal jelly.

iii   For sales

5. Grafting

Grafting is the transfer of a worker bee larva into an artificial queen cup for queen rearing. Grafting produces many high quality queens. The grafting is done by grafting tools.

6. Grafting Tool

Grafting tools is applied to transfer a worker larva into an artificial queen cup. The grafting tools is made of plastic or steel. Chinese grafting tools are also available in Indians market. A 'O' paint brush or a duck feather can also be used as the grafting tool.

7. Artificial queen cups

Artificial queen cups are moulded wax cups applied for queen rearing. 15-20 wax cups are fixed on a wooden bar called cell frame. The cell frame is fixed on the inner side of a comb frame called cell bar frame. The queen cup is protected by queen cell protector.

8. Acquiring bees

Obtaining honey bees for beekeeping is called acquiring be. The honey bees are acquired from the following sources:

1.   Catching a swarm.

2.   Purchasing package bees.

3.   Creation of a nucleus colony.

9. Uniting stocks

Combining the two honey bee colonies is called uniting stocks. It is the unification of the bee colony.

Uniting the stocks helps to save queenless colonies and weak colonies.

10. Splitting

Splitting is the division of a colony into two separate colonies. It reduces the colony strength.

11. Artificial Feeding

Feeding the bees with artificial feed is called artificial feeding. The artificial feeds include pollen, wheat, soyabean flour, sugar syrup, beat sugar syrup, sugar granules, etc. Feeds are given to feeders. The feeding may be inside feeding and outside feeding. Artificial feeding is given during lean period, queen rearing, requeening, acquiring bees, uniting stocks, etc.

12. Swarm Control

Swarming is basically the mass flight of a part of honey bee colony from a large colony to start a new colony. A swarm consists of a queen, workers and drones without a comb. Swarming produces a new colony, but weakens the parent colony. It is actually loss to the beekeepers. So swarming is controlled in modern beekeeping. Swarming may be controlled by the following methods:

1. Demaree method

2. Clipping of wings

3. Artificial swarming

13. Disease control

Honey bees may be infected by various pathogenic agents like fungi, bacteria, viruses, mites, etc. They cause diseases in honey bees. The diseases may be brood diseases or adult diseases. The following are the disease affection honey bees survival:

1. AFB (American Fowl Brood)

2. EFB (European Fowl Brood)

3. Shalk brood

4. Stone brood

5. Thai sac brood

6. Sac brood

7. Nosemosis

8. Acariosis

9. Varroosis

10. Clustering disease

11. Paralysis

12. Bee septicemia etc.

The diseases are prevented and treated by different prescribed medicines in modern beekeeping. In words of Albert Einstein- if bees dies or disappear, humanity can not survive more than four years. If we save the bees, we save the world. If there is no bees than there is no food. It is because 75% agricultural crops are pollinated by bees.

14. Control of Robbing

When there is scarcity of ford in the vicinity of the colony, its members attack other colonies and forcefully take out honey from them. This process is called robbing. Robbing is the thieving and stealing of honey from other hives. It occurs during dearth period. Robbing kills a colony and damages the comb. Robbing is prevented by reducing the size of the entrance and by keeping antirobbing screen.

15. Smoking

A smoker is used to calm boney bees. It helps to handle the honey bees and hives on various occasions like inspection of the hive, queen rearing, requeening, uniting stock, artificial feeding, honey extraction etc.

16. Clothings

Beekeeper is protected by veil, clothings and boots from the stinging of honey bees. It is for the safety of bee keeper from stinging of honey bees.

17. Honey Extraction

Honey is extracted by honey extractor without damaging the combs.

18. Honey processing

The extracted honey is processed through filtration and heating.

19. Royal jelly production

Royal jelly is the secretions of the worker bee for feeding larvae and queen. However, Royal jelly is exploited by man for medicinal purposes.

20. Queen selling

Superior quality queens are produced by queen rearing. The queen may be sold to beekeepers at a reasonable price. In modern method, first of all work is done to improve texture of the hives and during this race hive patterns were introduced in India.

Hive: The Newton model hive having 7 to 10 frames in the brood chamber with a shallow super has become most popular in the south, east and central India.

Longstroth's model containing 10 frames has been used as standard hive in Himachal Pradesh, Jammu and Kashmir and Punjab.

Movable hive is constructed which is capable of expansion or contractions according to the requirement of place, season and climatic conditions.

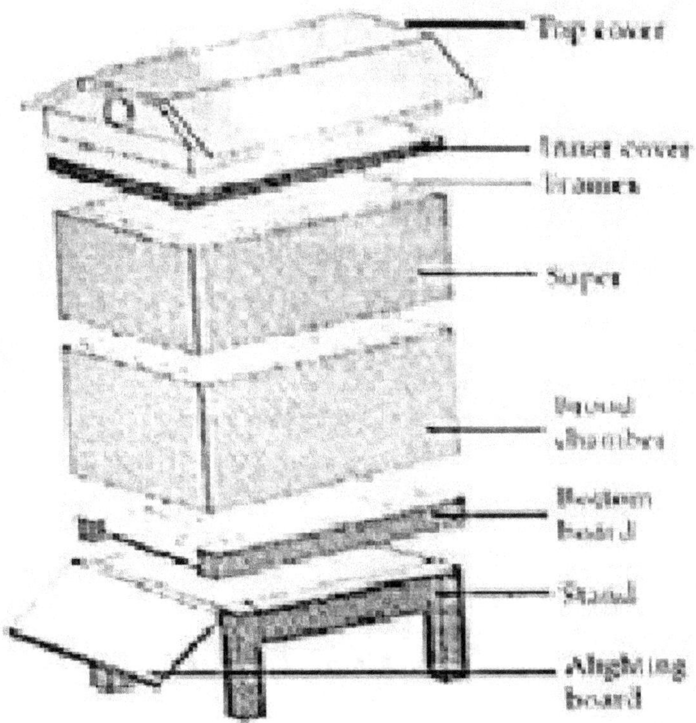

Figure 3.1- Langstroth's bee hive

Appliances for Modern Method

It consists of the following:

1. Typical movable hive.

2. Queen excluder.

3. Honey extractor.

4. Uncapping knife.

5. Other equipment.

1. Typical Movable Hive:

Now a days, a typical type of movable hive is made which may be expanded as well as contracted according to the new of the place, season and climatic condition.

An artificial movable hive is constructed by wooden box based on bee space theory. The size and number of frames are variable form hive to hive according to the need. A small space is enough to permit the entrance and exit of workers and drones but queen once placed in hive never comes outside the hive. The perforation size on zinc sheet is only of 0.375 cm but the thorax of the queen is 0.43 cm to 0.45, so the queen can never pass through the pore

Apiary (Apiculture)

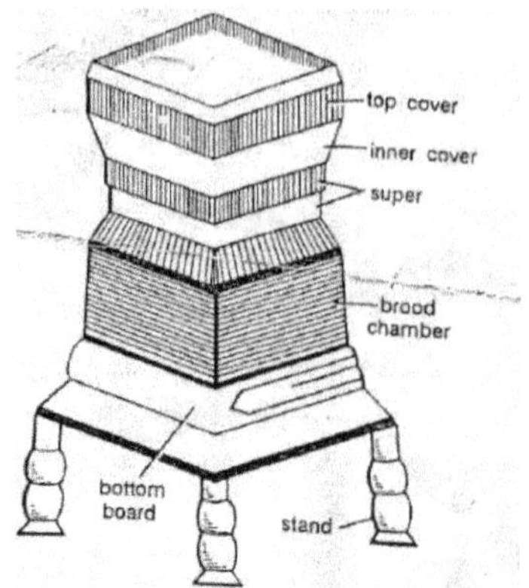

Fig. 3.2 (A) Typical movable hive.

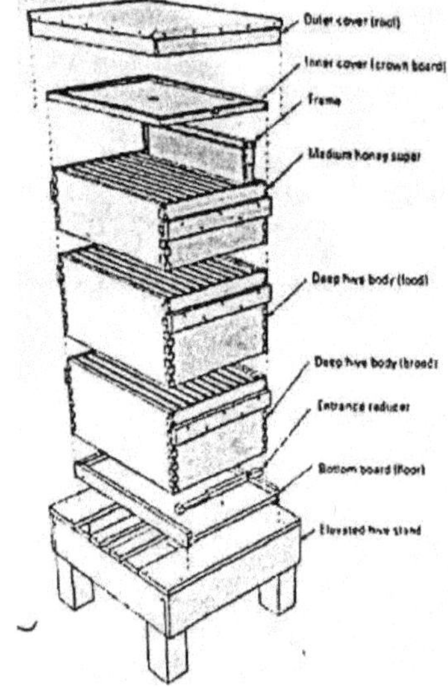

Fig. 3.2 (B) Modified movable hive

This typical hive consists of 6 parts as given below-

a)   Stand:-

It is the basal part of the hive on which the whole hive is constructed. The stands are adjusted to make slope for the hive. Due to this slope, rainwater comes down quickly.

b)   Bottom Board:-

It is situated above the stand and forms the proper base for the hive having two gates in the front position. One gate functions as an entrance while the other as exit.

c)   Brood Chamber:-

The bottom board carries the brood chamber, which is the most important part of the bee hive. It is large in size provided with 5 to 10 frames. In each frame a wax sheet bearing hexagonal frames is held up by a couple of wires in a vertical position. Along with the margin of every hexagonal mark, the bees start making will and ultimately the cell, are constructed.

Here every sheet of the wax is known as comb foundation, which attracts the bees and provides the base for the comb preparation on both the sides. The frames are kept vertically in brood chamber, which is covered over by other frames having a wire meshing through which the workers can easily pass.

The comb foundation helps in obtaining a regular string worker brood cell comb, which can be used repeatedly. The Central Bee Research Station at Pune arranged the manufacture of a comb foundation mill, which manufactures, different cell sizes required in several regions of the country. The brood chamber is covered by another chamber known as super.

d) Super:-

It is also without cover and the base. Super is provided with many frames containing comb foundation to provide additional space for expansion of the hive.

e) Inner Cover:-

It is a wooden piece used for the covering of the super. It has many holes for proper ventilation.

f) Top Cover:-

It is meant for protecting the colony from rains. It is fitted with zinc sheet, which is plain and sloping.

2. Queen Excluder:-

It consists for wire gauze. The drone traps with individual wires placed 0.375 cm apart. It readily permits the workers to pass through it but keeps back the queen in the brood chamber.

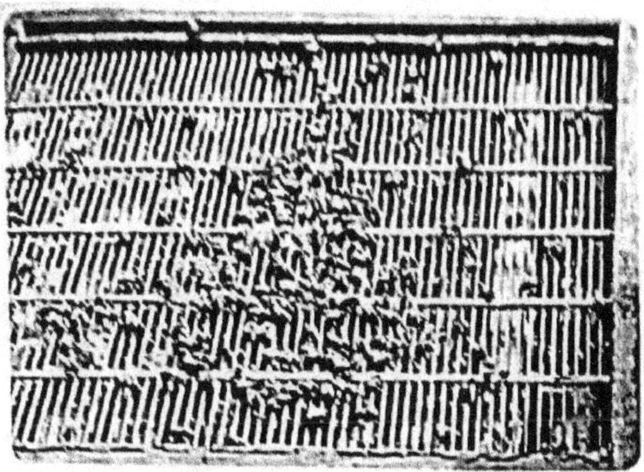

Fig 3.3- Queen excluder

3. Honey Extractor:-

It is used for the extraction of the honey from the comb and functions on principle of centrifugal force. When combs are centrifuged by this device, the pure honey is thrown out without any damage to the comb.

Fig 3.4- Honey extractor

4. Uncapping Knife:-

When all of the combs are filled with honey, they are sealed by capping with the wax. So, before such capped combs are placed in the honey extractor, the wax sealing has to be removed with the help of an uncapping knife heated by steam before use.

5. Other Equipment:-

Most of the useful, equipment for the successful management of the bee are locally manufactureed which are very cheap. As they are made locally, they may not be exactly similar to those made at other place. Thus, Indian standard Institute has standardized some very common equipment. For the production of uniform and interchangeable articles. Some materials like gum cages, gloves, net veil, bee net, brush etc. are required for easy and well planned handling of the bees.

Bee gloves are used by beekeepers for protecting their hands while inspecting the hives.

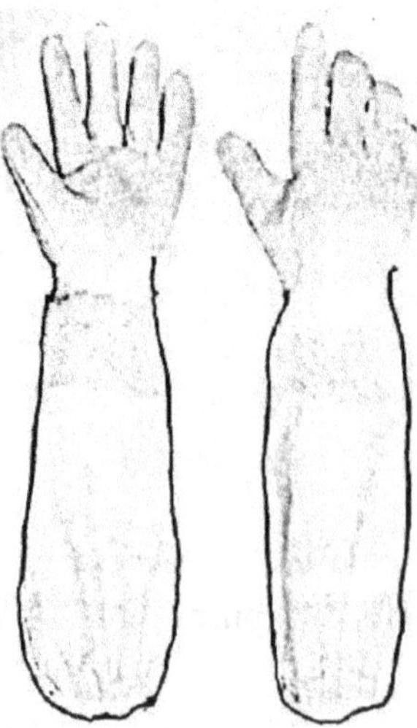

Fig 3.5- Bee gloves

Bee veil is a device made of fine netting to protect the beekeeper from the sting. Salient bee keeping equipments are shown in fig 3.6.

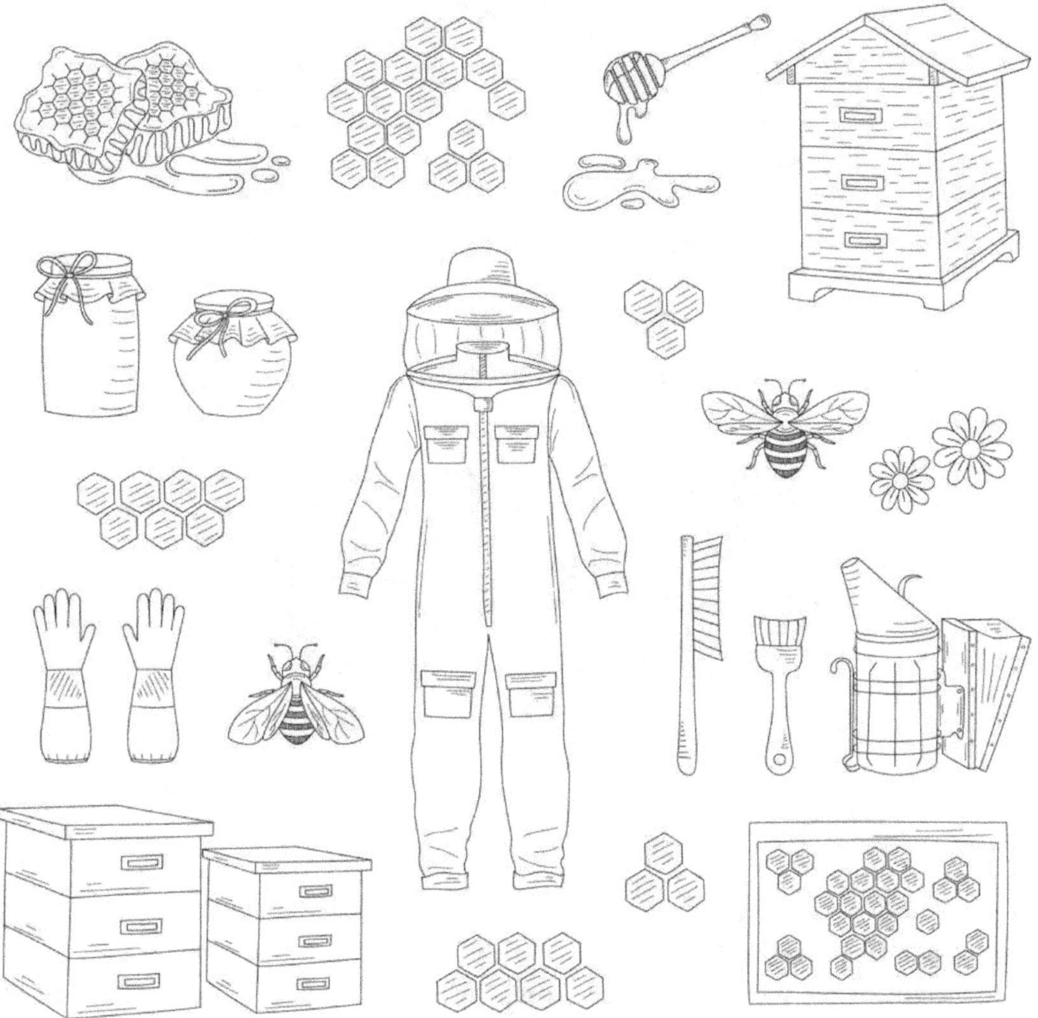

Fig 3.6- Bee Keeping Equipments

- Smoker is used to scare the bees during hive maintenance and honey collection by releasing smoke.

- Hive tool is a flat, narrow and ling piece of iron which helps in scraping excess propolis or wax from hive parts.

Fig 3.7- Flat hive tool

Fig 3.8- Use of Flat hive tools

Bee brush is a large brush often employed to brush off bees from Honey combs particularly at the time of Extraction.

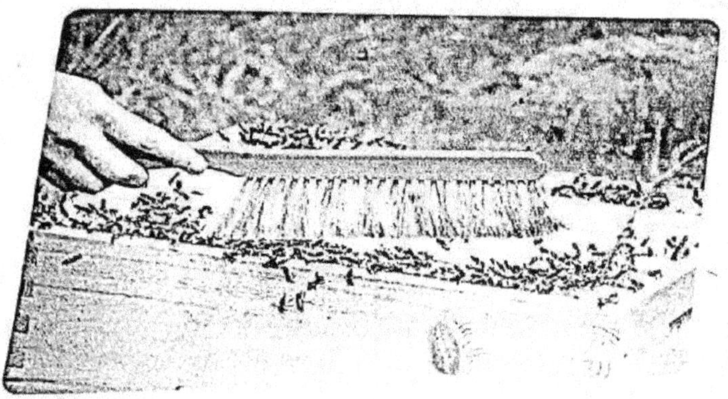

Fig 3.9- Use of bee brush

Advantages of Modern Method

In the modern method of bee keeping, there are several advantages which encourage the well planned bee keeping.

1. A proper watch on the activities of the bees can be made easy.

2. A strong colony can be developed by providing sugar, syrup, pollen substances etc. to honey bees.

3. Swarming of bees is checked by applying modern hive.

4. The same hive is used again and again so the workers pay their attention more for the honey and not for the hive formation.

5. Unfavorable climatic conditions, the hive can be transferred from one place to the other for the protection of the bees.

6. Comb can be protected from the enemies.

7. Pure honey in large quantity can be obtained.

Precautions: For the proper management of bee keeping programme, following precautions should be taken:

1. The hive should not be kept more than half a mile away from the place from where the bees have to collect the nectar and the pollen.

2. People must know about the bee keeping.

3. The boxes must be kept under shade at cool places.

4. Industry should be near the road for proper transport facilities.

5. Freshwater reservoir should be near the hive.

6. Good flora for the collection of pollen and nectar should be in the vicinity.

Salient Advantages of Modern Beekeeping (Hives)

1. Comb Foundation sheet may be further used.

2. The combs can be further reused.

3. The complete activities of the honey bees is in the knowledge of the beekeeper.

4. Artificial feeding can be provided.

5. Requeening is possible.

6. Queen rearing produces high quality queens.

7. Absconding is checked.

8. Robbing can be checked.

9. Swarming is controlled.

10. Bees get protection from enemies and weather.

11. Bee may be managed for pollination.

12. Colony splitting is possible.

13. Uniting stocks is possible.

14. Diseases can be controlled and treated.

15. Honey can be extracted by honey extractor.

16. Honey is pure and can be stored for ling period.

Advantages of Movable Frame Hives

The following are the salient advantages of movable bee hive:

1. Such hives are easier to construct because they have fewer areas where critical dimensions are important.

2. These can be easily made of materials which are readily available to the small farmers. Thus, they are more economical than langstroth- type hives.

3. These offer a cheap and intermediate alternative to beekeeping for bee-killers and bee-havers who are using fixed-comb hives.

4. There do not require foundation to guide the construction of comb within the frame to achieve their optimum returns.

5. Beeswax production is relatively high.

6. Honey can be harvested from new comb. Thus higher quality honey can be produced.

7. The top bars can be constructed so that they meet, leaving no openings along the top of the hive. This makes it easier to work with more defensive strains of bees.

Disadvantages of Movable Frame Hives

The following are the salient disadvantages of movable frame hive

1. The combs are attached only to the top bars, hence, it is difficult to move the colonies without breaking the combs. Also, care must be taken when removing combs and inspecting them.

2. Since the combs are attached to the top part of the hive, the colony can only expand in a horizontal plane. This somewhat limits the expansion of the brood nest, as the natural tendency of bees is to increase the brood nest in an upward direction (vertically).

Important Question:

1. Give an account of salient aspects of modern bee keeping.
2. Describe appliances for modern method of bee keeping.
3. Give an account of advantages and disadvantages of movable frame hive.
4. Write short notes on the following:
    A. Salient aspects of modern bee keeping
    B. Appliances for modern method of bee keeping
    C. Advantages of movable frame hive
    D. Disadvantages of movable frame hive
    E. Components of movable hives

UNIT - 4

Diseases Of Honeybee And Control Measures

Synopsis- (Adult Bee diseases (*Apis iridescent* Virus disease; Nesoma Protozoan disease; Acarosis; Varroasis; Amoeban diseases) Brood diseases of Honey bee (American foul brood; European Foul brood; Fungal brood and viral diseases).

Introduction

For high production of apiculture products, it is important to known about the diseases of honey bee. It is because the infected honey bee can not produce the honey or its products upto the marks. Most of the diseases in honey bees are of communicable nature, hence most of the members of colony may be affected within short period of time. Among various diseases of honey bees, viral diseases are highly dangerous and bring mortality soon.

The diseases of honeybees may be grouped into brood disease and adult disease.

The following are the adult diseases that have been observed in honeybees-

1. Apis iridescent virus disease
2. Nosema protozoan disease
3. Acarosis
4. Varroasis
5. Amoeban disease

Adult Bee Diseases:-

1. The diseases of adult bees are caused by protozoans which are single celled animals and form spores or cysts, damaging the internal system of the bees.
2. They multiply by sexual or asexual methods.
3. Their infection reduces vitality of bees, and shortens their life span and fecundity.
4. Protozoans are perfect parasites as they do not kill the host immediately.
5. These diseases are difficult to diagnose, though inability to fly, unhooked wings and dysentery can be treated as general symptoms of an unhealthy bee.
6. Microscopic examination is often necessary for a definite diagnosis. Salient diseases adult bee are the following:

1. Apis Iridescent Virus Disease

Symptoms:- The following are the symptoms of this disease:

1. Reduced egg laying / brood rearing.
2. Bees become sluggish and make cluster.

3. The infected bees crawl on the ground.

4. The affected bees cease foraging, reduce honey collection, and the colony face starvation and starts dwindling.

5. Illuminated body tissue can be seen under microscope looking bluish / greenish.

Fig 4.1 – Honey bee infected with Apis iridescent virus.

☐ Mode of Infection:-

Disease is caused by iridovirus. Its infection is serious during hot/ dearth seasons. It is transmitted by nurse bees through glandular secretion (food).

☐ Management of Viral Diseases:-

1. For viral pathogens, there is no chemical control
2. Affected colonies should be isolated beyond their flight range.
3. Proper ventilation should be provided to reduce humidity.
4. Cage the queen for a week and then requeen.
5. Use sterilized equipments / combs. Check robing, drifting and swarming.
6. Provide supplement feeding.
7. Undertake selective breeding for natural resistance.

Note: Paralysis in adult honey bees may occur after infection with virus. There appears trembling in the bee body. Affected honey bees stop flying and opt crawling. There is no satisfactory therapy against this disease and only preventive measures like regular cleaning of hive and nutritive food can protect them.

2. Nosema Disease

1. This disease is caused by *Nosema apis* (Zander), a protozoan of class- Sporozoa.

2. It is disease of adult bees. The parasite attacks the epithelium of mid gut causing loose motion and abdominal swelling. The parasite comes out faces and further produces infection in workers. It occurs through mouth during cleaning of the hive by worker bees and again parasites approach the GIT (gastro-intestinal tract) of the workers. The infection is enhanced gradually. This disease is common in cold climate regions but to some extent it also affects the bees during winter in the hot regions.

3. It parasitizes all the castes, but workers are the main target.

4. Their spores germinate in the venticulus of the host.

5. Pathogen multiplies in epithelial cells. Single affected bee may contain 180 million spores.

6. Infection spreads through ingestion of fecal matter with contaminated food. Hind gut is inflamed by protozoa.

Fig 4.2- Nosema disease of honey bee.

☐ **Symptoms-** The following are the symptoms of Nosema Disease:

1. Bees start foraging at younger age.

2. Bees feel fatigue, and are less able to fly and fall down during their return journey.

3. Abdomen is distended with fecal matter.

4. Body hairs are lost and bees become shiny.

5. Mid intestine is swollen and if dissected, shows dull greyish white contents. Bees soil the hive during entrance.

☐ **Management:-**
1. Provide fresh running water. Drain off stagnant water from the apiary.

2. Provide upward ventilation to reduce humidity.

3. Feed fumagillin in concentrated syrup. It inhibits DNA replication of the pathogen.

4. Disinfect the empty hives with ethylene oxide or acetic acid fumigation @ 120 ml/ hive.

3. **Acarosis (The Honey Bees Tracheal Mite)-** It is also called "Isle of Weight" disease.

☐ **Causal Agent:-**
Honey bee tracheal mite, *Acarapis woodi*, is a small parasitic mite. It affects mostly the trachea and body fluid. When it reaches the trachea of the honey bees, starts laying eggs over there. Presence of numerous eggs causes respiration to cease and finally honey bees die. The wings of the affected bees open to take a "k" shape.

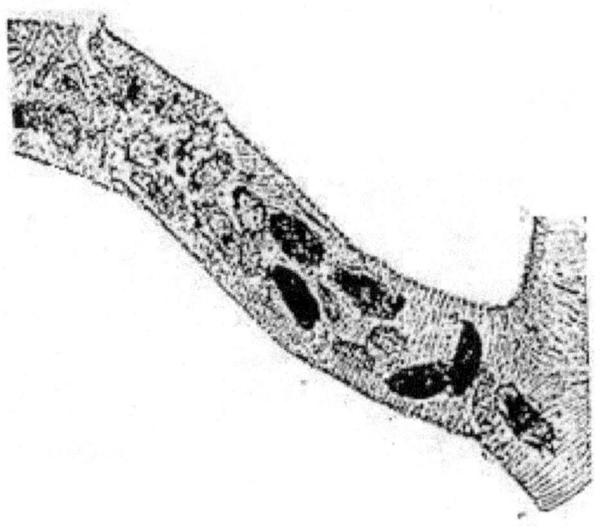

Fig 4.3 – Tracheal mites infecting honey bee

☐ **Nature of damage**

1. It infects worker, drone and queen honey bees. Mites live and reproduce in trachea.

2. They pierce the tracheal tube walls and feed on the hemolymph of the bees.

3. Feeding on blood and depositing their faeces in the passage.

☐ **Stage of Infection:- adult**

☐ **Place of Infection:- Trachea and body fluid**

☐ **Management:-**

1. Use of grease patties (typically made from 1 part vegetable shortenings mixed with 3-4 parts powdered sugar) placed on top bars of the hive. Menthol allowed to vaporize from crystal form or mixed into the grease patties.

2. Use of resistant hybrid bees known as Buckfast bee, developed by Brother Adam at the Buckfast Abbey.

3. Cotton soaked in Methyl salcilate and placed under the hive in flat perforated lid.

4. Destruction of affected colony.

5. Smoke fumigation with Chlorobenzilate.

☐ **Time of Treatment-** Spring and early summer.

4. Varroasis (The Varroa Mite)

☐ **Causal Agent:-** *Asiatic varroa mite, Varroa destructor.*

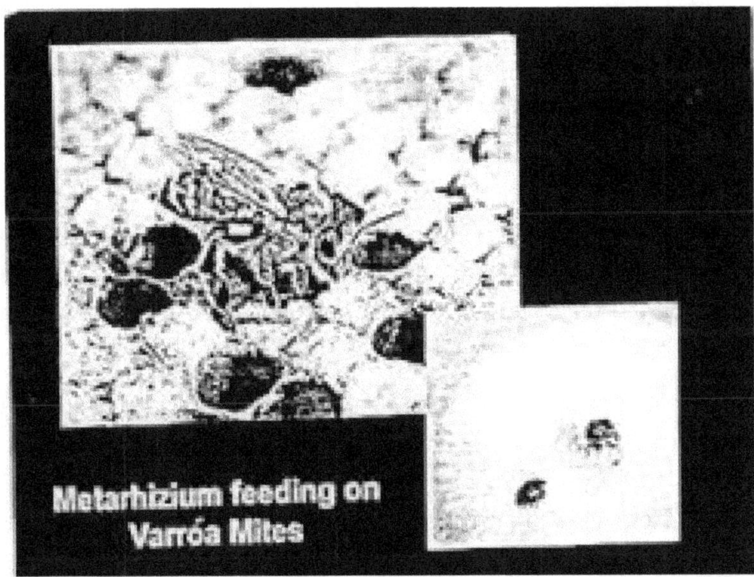

Fig 4.4 Metarhizium feeding on Varroa Mites

☐ **Symptoms:-** The following are the symptoms of Varroasis (The Varroa Mite)

1. Varroa reproduce on honey bee pupae and feed on bee hemolymph.

2. Varroa are also known to carry and vector bee viruses that are particularly damaging to the bees.

3. Varroa infestations can cause irreversible damage to honey bees that can lead to honey bee colony losses.

☐ **Stage of Infection-** Larval stage

☐ **Place of Infection-** Body and body fluid i.e. haemolymph.

Management:- The following measures should be taken for the management of this disease:

1. Apivar:- Apivar is effective against varroa mites, Apistan-resistant varroa mites, and Checkmite+ resistant varroa mites. Using 65% formic acid.

2. Mite Away Quick Strip (MAQS) – MAQS is a 7-day, single application mite control product registered for use against varroa and tracheal mites.

3. **Time of control** – Spring and early summer.

5. Amoebian Disease

1. It is caused by Malpighamoeba mellifecae.

2. This infection is caused by ingesting the cysts along with contaminated food.

3. Cysts germinate, amoebae migrate to malpighian tubules and feed on cell contents.

4. Cysts accumulate in the mid-gut / rectum.

5. Cysts are shed in the intestine and are excreted out with the fecal matter.

6. Peak infestation occurs during April-May.

☐ **Management:-**

Ensure proper hygienic conditions. Scrap off the bottom board and disinfect it with 2% carbolic acid. Disinfection of hive and equipments with acetic acid is also helpful.

ii. Brood Disease-

Larvae are infected by some fungi like *Pericystic apis* and *Aspergillus flavus*. Larval body is filled with the growing mycelia (hyphae) of these parasites and larvae are killed. This disease is incurable. The preventive measures like cleaning of hives, control of moisture, destruction of affected hives etc., should be taken to avoid infection.

In the same way, larvae are also infected by *Bacillus larvae* and *Streptococcus apis* bacteria and killed ultimately. Infected larvae give foul smell. Infection occurs through spores which enter the body with honey. This disease spreads rapidly and results into mass killing of honey bees. This disease is most common in cold regions and chances of infection in hot regions are less. There is no remedy for this disease. The infected hives should be burnt without delay.

The following are the brood diseases of honey bee.

1. American foul brood

2. European foul brood

3. Fungal brood diseases- It is of two types namely

   A Chalk brood

   B Stone brood

   4. Viral diseases (Thai sac brood)

1. American foul brood disease (AFB disease):-

This is the most destructive microbial disease in temperate and sub-tropical region all over the world.

☐ **Cause:-** Spore forming bacterium, *Bacillus larvae* only affect brood.

☐ **Symptoms:-**

The symptoms of AFB disease include: Initial stage, isolated capped cells from which brood has not emerged can be seen on the comb. Caps of these brood cells are darker than the caps of healthy cells. The infected brood die at prepupal or late larval stage. The dead brood is dull white in color, but gradually changes to light brown, coffee brown and finally dark brown or almost black. Dead pupae appear and sunken brood.

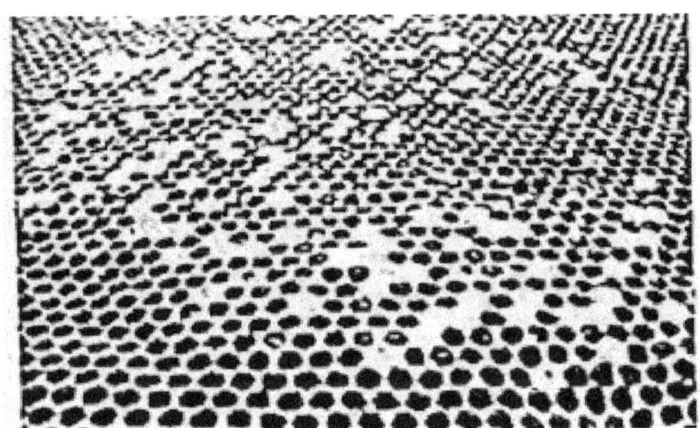

Fig 4.5 American foul brood Scale

☐ **Diagnostic procedure:-**

1. The simple test for AFB is the "Stretch test".

2. A match stick or tooth stick is inserted into the body of the decayed larva and then, gently and slowly, withdrawn.

3. If the disease is present, the dead larval content adhere to the tip of the stick, stretching up to 2.5 cm before breaking and snapping back in somewhat elastic way.

☐ **Treatment:-** The treatment of AFB is the follows:

1. Feeding of sodium sulfathiazole @0.1 g/liter in sugar surup.

2. Feeding oxytetracycline (Terramycin) 0.25 to 0.4g in 5 liter sugar syrup.

3. Feeding streptomycine in sugar syrup @0.05 – 0.15 g/liter.

4. Dust Terramycin (TM50) in powdered sugar (1.20) @ 4 tea spoon full on top bars of the brood frames.

5. Chemotherapy has no effect on spores that contaminate the equipment.

6. Chemotherapy is no advisable for low disease indication.

2. European Foul Brood Disease (EFB disease):-

In India, except Maharastra, EFB has been recorded so far in *Apis mellifera* colonies.

☐ **Cause:-** This disease is caused by Pathogenic gram negative bacterium *Melissococcus pluton*.

☐ **Symptoms:-** The following are the symptoms of EFB disease:

i. Honeybee larvae killed by EFB are in younger stages than those killed by AFB.

ii. The diseased larvae die when they are 4 to 5 days old, or in the coiled stage.

iii. The color of the larvae changed from shiny white to pale yellow and then brown, as it decays.

iv. Another symptom of EFB is that most of the affected larvae die before their cells are capped.

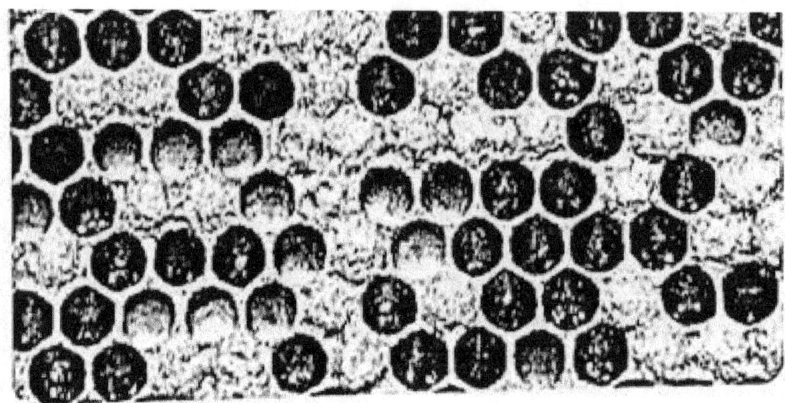

Fig 4.6 – European Foul Brood Disease

☐ **Treatment:-**

1. Not required if the infection is low.

2. Sterilization of combs and honey could be done with formaldehyde and acetic acid.

3. 0.5 or 1 g oxytetracycline (tetramycin) dissolved in 500 ml of concentrated sucrose syrup sprinkled on bee clusters gave good results.

3. Fungal Brood Diseases:-

It is of two types –

(i)   Chalk brood and

(ii) Stone brood

(i)   Chalk brood:-

☐ **Cause:-** The following is the cause of chalk brood disease

1. Causative agent of this disease is *Ascophaer apis.*

2. It is a heterothallic fungus.

3. Colonies rarely die from this disease, but in some cases honey yield may be reduced.

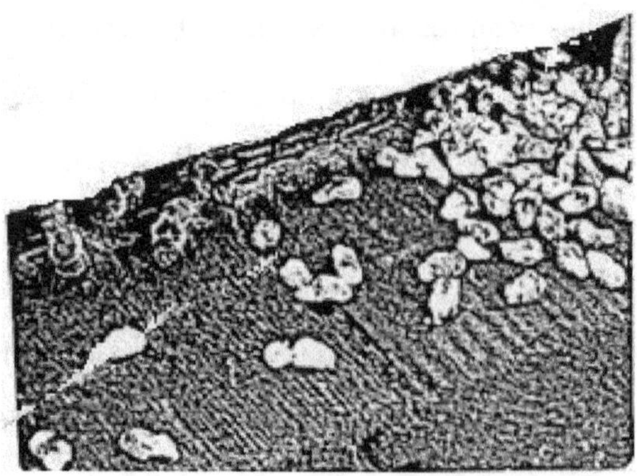

Fig 4.7- Chalk brood of honey bee.

☐ **Symptoms:-** The following are the symptoms of chalk brood disease:

1. It affects honeybee larvae or brood.

2. Chalk brood is usually more common on the edges of the brood chamber.

3. It is most common in drone brood.

4. Initially, the dead larvae are swollen to the size of the cell and covered with the whitish mycelia of the fungus.

5. Subsequently, the dead larvae become mummified, hard, shrunken and chalk like in appearance, hence the name is chalk brood disease.

☐ **Treatment:-** The following are the treatments of the Chalk Brood disease

1. Equiment's should be sterilized using formalin or carbolic acid.

2. 0.7% of thymol has been reported to prevent the growth of fungus.

3. Amphotericin B, sorbic acid and sodium propionate fed to bees in pollen-sugar patties controlled infection within 7 days without affecting bees.

4. Nistapi R controls more than 98% of the infection in affected colonies.

5. Increasing ventilation.

6. Removing "mummies" from bottom board and around the entrance.

7. Destroying combs containing large numbers of mummies.

(ii) Stone Brood:- It is another fungal disease.

☐ **Cause:-** It is caused by *Aspergillus niger, Aspergillus fumigates and Aspergills flavus.*

Fig 4.8. Stone Brood

☐ **Symptoms:-** The following are the symptoms of stone brood disease:

1. Mummification of brood.

2. Larvae and pupae that are infected with *A. flavus* turn green in contrast to white or black in chalk brood.

3. The green growth is powdery and be readily seen with unaided eye.

4. Fungus spores are found most abundantly near the head of the affected larvae and pupae.

5. Stone brood diseased larvae are solid mummies and not sponge like as in chalk brood.

6. **Viral Disease:-** So far, the world over, 18 viruses have been found to infect honeybees. Thai sac brood virus and *Apis iridescent* virus have been found to cause heavy losses to bee industry in India.

(i). Thai sac brood:- It is viral disease.

☐ **Cause:-** It is caused by Thai Sac Brood Virus (TSBV) which primarily infects the larvae of *Apis cerana* i.e.*Apis indica* and is closely related to sac brood virus (SBV) which infects the *Apis mellifera*, but the two are not identical.

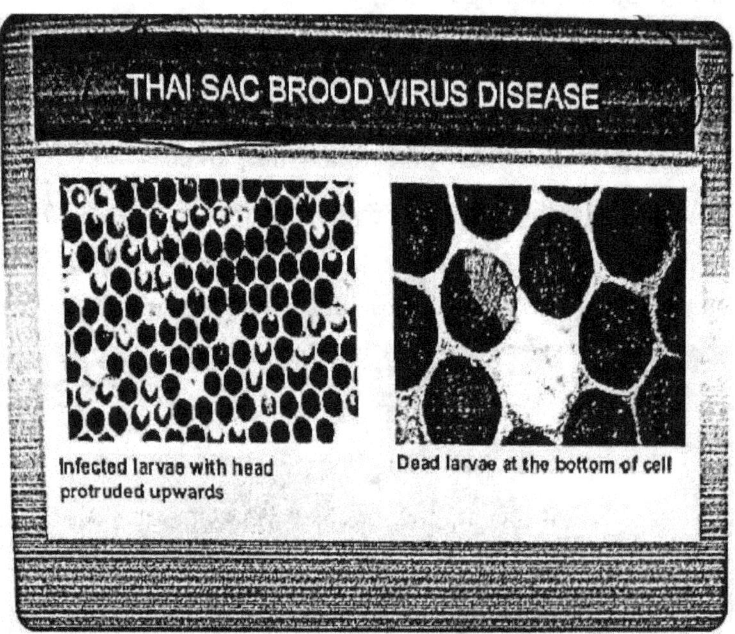

Fig 4.9 Thai sac brood virus disease

☐ **Symptoms:-** The following are the salient symptoms of the Thai Sac Brood Virus disease:

1. This results in the death of the larva in the uncapped cells.

2. Diseased colonies show irregularly capped brood with sunken and faded caps.

3. The head of the dead larva is turned up partly across the cell opening.

4. Infected larva turn pale yellow and finally brownish when dead.

Besides, TSB, Kashmir bee virus, Black queen cell virus and cloudy wing virus also caused diseases in early developmental stages of honey bees.

A. Kashmir bee virus:- Kashmir bee virus is related to the preceding viruses. Recently discovered, it is currently only positively identifiable by a laboratory test.

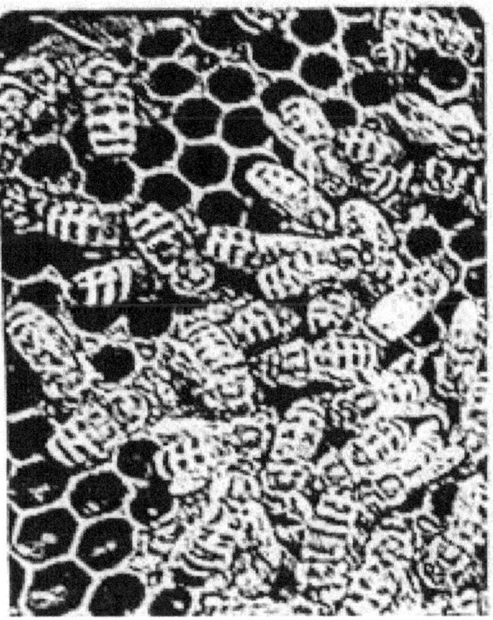

Fig 4.10 Kashmir bee virus

A. **Black queen cell virus:-** Black queen cell virus causes the queen larva to turn black and die. It is thought to be associated with Nosema.

B. **Cloudy wing virus:-** Cloudy wings virus is little-studied, small isohedral virus commonly found in honey bees, especially in collapsing colonies infested by *Varroa destructor*, providing circumstantial evidence that the mite may act as a vector.

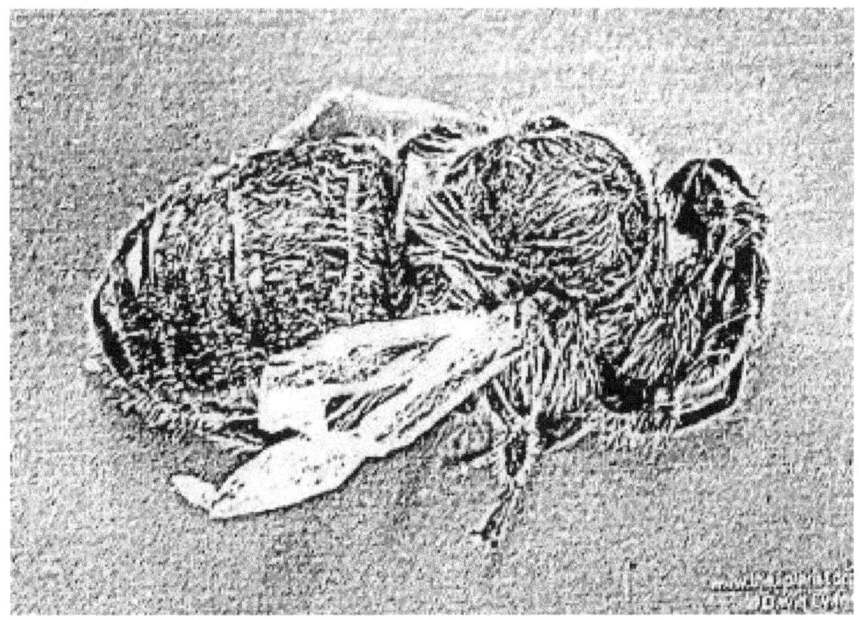

Fig 4.11 – Honey bee infected with CWV.

☐ **Control measurements:-** The following are the control measures for above three viruses:

1. Use of antibiotics such as Rifampin, Leavamisol, Amentidine along with vitamin B complex fed to honeybee colonies @ 250mg/4 lit of sugary syrup at weekly intervals.

2. Avoiding of overcrowding.

Important Questions:

3. Give an account of various fungal diseases in honey bees.
4. Describe viral diseases in honey bees.
5. Give an account of Protozoan diseases in honey bee.
6. Write short notes on the following:
   A. Thai Sac Brood Virus
   B. Kashmir Bee Virus
   C. Black Queen Cell Virus
   D. Cloudy Wing Virus
   E. Acarosis
   F. Apis Iridescent Virus Disease
   G. Nosema Disease
   H. Chalk Brood Disease
   I. Stone Brood Disease
   J. American Foul Brood Disease
   K. European Foul Brood Disease

UNIT - 5

Bee Enemies

Synopsis - (Insects (Ants, wasps, moths); Vertebrates and microbes.

Introduction

In India, honey bee keeping industry is affected more severely by the natural enemies of the bees than the parasites. Following are some salient natural enemies of honey bees-

Insects

1. Ants-

There are several predatory species of ants like red ant *Dorylus labiatus*, black ant, *Dorylus orientalis*, carpenter ant *Componotus compressor* etc., which feed on eggs and larva of honey bees and pollens causing heavy damage to apiary. So the nearby nests of ants should be destroyed by applying repellants like corrosive sublimate.

2. Wasps-

Many wasp species like *Philanthus ramakrishnae*, *Philanthus trianglum*, *Vespa orientalis*, *V. magnifera* etc. attack honey bees and feed upon them. There for the nearby wasp nests should be destroyed.

3. Moths

There are many species of moths also which cause great damage to honey bees. Death's head hawk moth, *Acherontia styx* is a large sized moth causing damage to honey and hive during night. It sucks the honey. Another giant moth *Galleria mellonella* also called greater wax moth which lays eggs on honey bee hive. Larvae coming out of eggs pierce the hive and penetrate inside to feed on pollens.

These moth larvae secrete silk and clog the honey cells. If required, repellent like para dichlorobenzene should be used to drive away these moths.

4. Beetles- Beetle *Platybolium alvearium* of the family Tenebrionidae takes out honey from the hive. Para dichlorobenzene should be used in this case to protect the honey bees.

5. Varrowa- It is a mite which sticks with various body parts of honey bees and sucks its food. Honey bees can be protected by gentle sulphur dusting over hives.

6. Lizards and Frogs- These are insectivorous vertebrates which feed upon various stages of honey bees. To protect the honey bees, hives should be kept on stands.

7. Birds- Crows, merops, neelkanth or Indian roller (*Caracium bengalensis*)etc., also feed upon honey bees. Attention should be given to check them to approach bee hive.

8. Mammals- Spiny rat hedgehog, mice, bear and many other mammals are enemies of honey bees. Amongst these, bears harm much to bee high because they prefer taking honey.

Honeybees are attacked by various pests, predators and other enemies. The major bee enemies are wax moths, wasps, birds, ants, hive beetles, mice and bear, which destroy the raised combs, hives and hive parts, catch and kill bees, colony development, eat away the food reserves and cause nuisance to the bees.

The enemies of honey bee may broadly be categorized in following categories.

1. Arthropods especially insects

2. Vertebrates

3. Microbes

1. Arthropods especially insects- The following are the well known Arthropods enemies of honey bee:

☐ **Beetle:-** There are several different beetles found living in honey bee colonies. Most of them are harmless and feed on pollen or honey.

A Small hive beetle- Name- Aethinatumida (order- Coleoptera, family- Nitidulidae)

The beetles and their larvae form eating canals and destroy the cell caps, and the honey starts to ferment. The beetles larvae and faeces also change the colour and taste of the honey and the combs appear mucilaginous.

Control-The best way to protect against an infestation of the small hive beetle is to keep strong colonies and to remove those that are weak from an apiary. The removed honey combs should be centrifugally extracted one to two days after harvesting the honey.

B Ants:-
Ants are among the most common predators of honey bees in tropical and subtropical Asia. They are highly social insects and attack the hives and in mass, taking virtually everything in them-dead or alive adult bees, the brood and honey.Weak colonies will sometimes abscond, which is also the defence of *A. cerana* i.e. *A. Indiaca* against frequent ant invasions.The weaver ant (*Oecophyllas maragdina*), the black ant (*Monomorium indicum*), *Monomorium destructor* and *Oligomyrmex* spp., *Dorylus* spp., the fore ants (*Solenopsis* spp.) and *Formica* spp. are salient ants that cause harm to the apiary directly or indirectly

Control- Apiarist must search thoroughly for the ants nests in the vicinity of the apiaries and when found should to destroy them by burning.

Place the hives on stands supported by posts 30 to 50 cm high and to coat the posts with used engine oil or grease, or place the stand in a container with oil or water.

C Wasps and Hornets:-Nearly all countries in Asia report reveals that wasps and hornets as common enemies of honey bees.
- Causal organism- *Vespa* spp., which attacks on *Apis cerana* i.e. *A. indica* and *A. mellifera*.

- The affected bees colonies undergoes absconding.
- In temperate Asia, hornet attacks on apiaries during September-October.
- In tropical countries the most serious wasp invasions takes place during the monsoon season.
- The hornets invade the hive itself, the honey and brood nest and the wasps carry away any surplus brood to their nest.

Control:- The following are the control measures of wasps and hornets:
- Destroy the wasp nest either mechanically or by using fumigant like Aluminium phosphide or Calcium cyanide.
- Maintain strong colony.
- Use of wasp trap- Ripened jack fruit with 2 grams of Furadan.
- Covering the colony with coconut branches.

D Wax Moths:- The greater wax moth (*Galleria mellonella*) is well known enemy of apiary
- The wax moth is a major pest of *A. indica*, often causing colonies to abscond.
- The adult female enters the hive at night, through the entrance or breaks in the walls, deposits her eggs directly into the combs or in narrow crevices,
- The newly hatched *Galleria* larvae feed on honey, wax and Pollen, and then burrow into Pollen storage cells or the outer edge of cell walls, later extending their tunnels to the midrib of the comb as they grow.
- The larvae also take on bee brood when there is shortage of food.

Control:- The following are the control measures for wax moths:
- Frequent examination of the hive.
- Cleaning all the crevices and removing all the debris.
- The excess combs in the hive not covered by the bees are removed and stored after the fumigation with methyl bromide.
- In the stored rooms the spare combs should be stored in tightly closed containers.
- Store the empty combs at low temperature (0-10° C) either permanently or for 5 hours.
- All the stages of wax moth are destroyed at low temperature.

E The Lesser Way Moth (*Achroia grisella*):-
- Generally smaller than the greater wax moth.
- Adult *Achroia grisella* are silver- grey in colour, with a distinct yellow head. The insect is quite small, with a slender body.
- The life- span of the adult female is about seven days.
- They usually infect weak Honey Bee colonies.
- The larvae prefer to feed on dark comb, with Pollen or brood cells.
- They are often found on the bottom board among the wax debris.

Control:- The following are the control measures for the Lesser Way Moth:
- Remove the infected comb and destroy either by burning or digging inside the soil.
- Unite the weaker colonies.
- Clean the bottom board at fortnight interval.

- Adequate food supply to maintain a strong colony (sugar feeding).
- Always maintain healthy comb.
- While staking the super chamber place 1/3rd Tablet of Aluminium phosphide for every cubic ft.

Non-Insect Enemies

I Spider- Pseudo scorpion (*Ellingsenius* sp.), Mites and Endo Parasitic Mite of genus *Acarapis* are the salient non-insect enemies of the apiary. They cause acarine disease in the honey bee. Both nymphs and adults feed the haemolymph from the young worker bee by entering through the spiracles.

☐ **Symptoms:-** Following are the symptoms of Spider and Endo-parasitic mite infection on honey bees:

- Presence of crawler bee in front of hives.
- Fore wing and Hind wing get separated resembling k shaped wing.
- Yellow dropping due to dysentery,
- Distended and shinning abdomen.

☐ **Control:-** The following are the control measures for non-insect enemies:
- Smoke fumigation with chlorobenzilate one strip/hive.

II Exo - Parasitic Mite- Causal Organism:- *Tropilaelaps clareae* commonly called as Rust red mite attacking both brood and adult bees. It spreads through rock bee.

☐ **Symptoms:-** The following are the symptoms of attack of Exo-parasitic mite

- Irregular brood pattern.
- Dead malformed larvae/ pupae/adults.

☐ **Control:-** The following is the control measure for Exo-parasitic mite:
- Fumigation with 85% Formic acid 5 ml./ hive daily for 21 days.

1. Vertebrates:- The following vertebrates are known to harm the apiary directly or indirectly:

A. Amphibians- Toads including *Bufo melanostictus* and *Koloula pulchra* and frogs including *Rana limnocharis* and *Rana tigrina* are well known enemies of the apiary.

☐ **Control:-** The following is the control measure for Amphibians as enemies of honey bees:
- Placing the hive on stands 40 to 60 cm high is usually a sufficient protective measure from Amphibians.

A. Reptiles:- The following Reptiles harm apiary directly or indirectly:

Geckos, skinks and other lizards (*Calotes*) are among the most commonly found reptiles in tropical Asian jungles, woods, grasslands and urban areas that harm to apiary.

☐ **Control:-** Following are the control measures of Reptiles:
- Hives placed on stands that about are 40-60 cm high reasonably safe from reptiles attacking from ground.
- Coating the legs of the stands with used engine oil or grease may deter the reptiles from climbing up to the Hive entrance.

A. Birds:-

Birds that have been listed as attacking honey bees in Asia include bee-eaters (*Merops apiaster, Merops orientalis*), swifts (*Cypselus* sp., *Apus* sp.), drongos (*Dicurus* sp.) shrikes (*Lanius* sp.), woodpeakers (*Picus* sp.) and honey guides (*Indica toridae*).

The level of damage caused by apivorous birds varies. An attack by a single birds or by a few together rarely constitutes a serious problems, but when a large flock descends upon a few colonies or an apiary, a substantial decline in the worker population, in some or all the hives may be observed.

Control:-

Heavy predation by birds on apiary bees tends to occur at fixed periods (e.g., during the migration season of swifts). The most practical means of solving the problem is usually to avoid the birds through careful site selection and by temporary relocation of the apiaries.

D. Mammals:- The Rat, bear etc. are well known mammals causing harm directly or indirectly to apiary. They prey in colonies for honey and brood.

Control:-

- Electrified barbed-wire fences are often used where bears represent a common problem, shooting and trapping them are other possibilities but very temporary control measures.

1. **Microbial Diseases:-** It includes the following:

A. **Bacterial Diseases:-** It includes the following:

I American foulbrood disease (AFB):- It is caused by *Paenibacillus larvae.*
- It is the most widespread and destructive of the honey bee brood diseases.
- It affects queen, drone, and worker larvae alike.
- Place of infection- Gut

- **Symptoms-** The following are the symptoms of AFB:
Turn dark brown and later changes into sticky mass producing foul smell (Infected larvae darken and die) Dead pupae, irregular and sunken brood.

- Stage infected- Larvae

- **Management-** The following measures should be taken for management of AFB:
In hive, a completely use of antibiotics such as oxytetracycline hydrochloride (terramycin) and tylosin tartrate Dusting the combs with sulphathiazole powder Dipping the hive in hot paraffin wax or a 3% sodium hypochlorite solution (bleach). Burning of infested comb.

II European foul brood (EFB):-

- Causal Organism:- *Melissococcus plutonius, Bacillus pluton* (bacterium).

- Place of infection:- Mid-gut

- **Symptoms:-** The following are the symptoms of EFB:

- The diseased larvae turns yellow and then brown and the tracheal system becomes visible.

- Larvae dies in a coiled stage causing foul smell. Cells are poorly capped and mixed with normal cells.

- Stage Infected:- Larvae

- **Management:-** The following measures are taken for management of infection of EFB:
- Use of oxytetracycline hydrochloride.
- The "Shook Swarm" technique of bee husbandry can also be used to effectively control the disease.

A. Fungal Diseases:- Two fungal diseases are found in apiary- (a) Chalk brood disease and (b) Stone brood disease.

a. Chalk brood (Kalkbrut disease):-

- Causal Organism – *Ascosphaera* apis

- Place of Infection - Gut

- **Symptoms-** The following is the symptoms of Chalk brood disease:
- The fungus will consume the rest of the larva's body, causing it to appearwhite and 'chalky'.

- Stage Infected – Larvae

- **Management:-** The following measures should be taken for control of Chalk brood disease:
- Prevent the hive from wet spring.
- Transfer of healthy bees into another bee hive.
- Increase the ventilation through the hive.

a. Stone Brood (Steinbrut disease):-
- Causal Organism:-*Aspergillus fumigates, A.. flavus and A. niger.*

- Place of Infection:- Alimentary canal.

- **Symptoms:-** Following are the salient symptoms of stone brood disease:
- Dead larvae turn black and become difficult to crush, hence the name stone brood.
- Fungus erupts from the integument of the larva and forms a false skin and larvae are covered with powdery fungal spores.

- Stage Infected:- Larvae and adults

- **Management:-** The following is the control measure for stone brood disease:
- Sterilization of the hive with formaldehyde fumes.

B. Viral Diseases:- Thai Sac Brood:- It is a serious brood disease of *Apis cerana* (*indica*) caused by virus *Moratora* etatulus (Thai strain) during spring season.

- **Symptoms:-** Following are the symptoms of Thai sac brood disease in honey bees:
- Spotty brood appearance (pepper box);
- Capping tends to be darker, concave and punctured frequently.
- Dead larvae dry up in brood cell forming loose scale or sac like.

- Brood die in pre pupal stage but in unsealed stage.
- Dead larvae lie on their back, tip of head capsule turned upwards.

☐ **Management:-** The following are the control measures for Thai sac brood disease in honey bees:

- Destruction of frame.
- Maintenance of strong and vigorous colony.
- Sterilization of beekeeping equipments with KMnO4 @ 50g/lit water followed by hot water dipping.
- Fumigation of hive by formic acid (85%) @5ml/hive.
- Caging queen for 20 days for creating broodlessness condition.

Important Questions

1. Give an account of various insect enemies of honey bees.

2. Describe the Vertebrate enemy of the honey bees.

3. Give an account of the various microbes as enemies of honey bees.

4. Write short notes on the following:

   A. Ants as bee enemy
   B. Wasps as bee enemy
   C. Moths as bee enemy
   D. Vertebrates as bee enemy
   E. Microbes as bee enemy
   F. Bacterial disease in honey bee
   G. Fungal diseases in honey bee
   H. Viral diseases in honey bee

UNIT - 6

Honey Bees For Cross Pollination In Horticulture Garden

Synopsis- (Introduction; Modern methods in employing Honey bees for cross pollination; Qualities of Honey bees which make them good pollinators; Management of bees for pollination; Effects of bee pollination on crop; Natural and ornamental flowers).

Introduction:-

The process of transfer of pollen grains (microspores) of a flowering plant species from anther to the stigma of the plant of the same species through any agent or medium is called pollination. If the pollination is through insects e.g., honey bees, then such pollination is called Entomophilous pollination. Honey bees play very important role in the pollination of the horticulture angiospermic flora. The following flora which are not visited by the honey bees bear comparatively less fruits. Value of honey bees in pollination is 15-20 times higher than that of the honey and wax. Due to cross pollination, the yield of crop is increased many folds as shown in Table 6.1. In view of Albert Einstein, if we save the bees, we save the world. It is because about 75% of agricultural flora are pollinated by honey bees. Therefore, if we save the bees, we save the world.

Table 6.1- % increase in yield of different crops due to Bee Pollination

S. No.	Crops	% increase in yield due to pollination
1.	Mustard	43%
2.	Sunflowers	32-48%
3.	Cotton	17-19%
4.	Lucerne	90-112%
5.	Onion	93%
6.	Apple	44%
7.	Cardamom	21-37%

Pollination is the release of pollen grains (n) from pollen sac through any media to the stigma of flowers is called Pollination. Pollination may though various agencies like air, water, animals and insects. Pollination through air is called aeroponilous pollination; through water is called hydrophilous pollination; through animal is called zoophilous pollination and pollination through insects viz; honey bee is called Entomophilous pollination. Horticulture gardens have varieties of flowering (angiospermic) plants, hence bees have great importance for cross pollination.

Modern method in Employing Honey Bees for cross Pollination in Horticulture Garden

Cross pollination is transfer of pollen from one plant to stigma of another plant. Honey bees play most salient role in pollination, therefore, called as fast friend of farmers.

Honey bee as pollinators

Honey bee visits plants for its food, nectar and pollen. This floral fidelity of bees is due to their preference for nectars having sugar contents and pollens with higher nutritive values. Besides getting food for the bees as a result of their visit pollinate a number of 75% crops.

- Bee prove excellent pollinators because most of their life is spent collecting pollen, a source of protein that they feed to their developing offspring.
- The effectiveness of honey bees is due to their great number, their social life and their ability to pollinate a broad variety flowers in horticulture and agriculture.
- A colony may consist of 20,000- 80,000 bees, and they will normally be visiting flowers over a distance of two kilometers when they are collecting pollen and nectar.

Qualities of honeybees which make them good pollinators:-

Following points make bees good pollinator-

- Body covered with hairs and has structural adaptation for carrying nectar and pollen.
- Bees do not injure the plants.
- Adult and larva feed on nectar and pollen which is available in plenty.
- Considered as superior pollinators, since store pollen and nectar for future use.
- No diapauses is observed and needs pollen throughout the year.
- Body size and proboscis length is very much suitable for many crops.
- Pollinate wide variety of crops.
- Forage in extreme weather conditions also.

Management of bees for pollination:-

Following points are for the management of bees for pollination-

- Place hives very near the field source to save bee's energy.
- Migrate colonies near field at 10 per cent flowering
- Place colonies at 3/ha for Italian bee and 5/ha for Indian honey bee.
- The colonies should have 5 to 6 frame strength of bees, with sealed brood and young mated queen.
- Allow sufficient space for pollen and honey storage.

Effects of bee pollination on crop:-

Following are the effects of bee pollination on crops-

- It increases yield in terms of seed yield and fruit yield in many crops. Keeping this in view, one of the author (JPS) cites that "no crop without honey and no honey without crops" or "no agriculture without apiculture and no apiculture without agriculture".
- It improves quality of fruits and seeds.
- Bee pollination increases oil content of seeds in sunflower.
- Bee pollination is a must in some self incompatible crops for seed set.

Crops benefited by bee pollination:- The following group of different crops are benefitted by bee pollination-

1. Fruits and nuts:- Almond, apple, apricot, peach, strawberry, citrus and litchi.

2. Vegetable and vegetable seed crops:- Cabbage, cauliflower, carrot, coriander, cucumber, melon, onion, pumpkin, radish and turnip.

3. Oil seed crops:- Sunflower, niger, rape seed, mustard, safflower, gingelly.

4. Forage seed crops:- Lucerne, clover.

Scope of beekeeping for pollination in India:

Total area of bee dependant crops in India is around 50 million hectare. One hundred and fifty million colonies are needed to meet this, at the rate of 3 colonies per hectare. In India at present, there are only about 1.2 million colonies exist. Hence there is a wide scope for expansion of bee keeping for pollination in India.

Pollination by bees:- Following are the important crops which are pollinated by honey bees-

1. **Sunflower:-** It is a cross-pollinated crop. The pollen of the plant cannot fertilize ovary of same plant. Pollen source should be from different plant. Hence, honey bees acts as important agents for pollination in sunflower. In sunflower, yield increases even up to 600 percent due to bee pollination. It improves quality and quantity of seeds. Oil content also increases by 6.5 per cent in seeds. To achieve this it requires five strong *Apis indica* colonies or three *A. mellifera* colonies. Mostly irrigated crops are preferred by bees.

   2. **Cucurbitaceous vegetables:-** Cucurbits are monoecious with staminate and pistillate flowers in same plant. Due to bee pollination fruit set increases up to 30 to 100 percent.

   3. **Alfa lafa or Lucerne:-** These plants have tubular flowers with 5 petals joined at base. They posses one large standard petal, 2 smaller petals on sides and 2 keel petals holding staminal column. When bee sits on a keel petal, stamina column strikes against standard petal resulting in shattering of pollen. This is called tripping. Seed set occurs only if bee sits to trips the flowers.

   4. **Coriander:-** In coriander, yield increases up to 87 per cent due to bee pollination.

   5. **Cardamom:-** It is an important commercial crop depending on bees for pollination. Here yield increases up to 21 to 30 percent.

   6. **Gingelly:-** Another oilseed crop where bee pollination causes 25 percent increase in yield.

7. **Apple:-** In apple seed set occurs only if it is pollinated by bees. Fruit is formed only around the seeds. If improper seeds set occurs fruit shape is not proper resulting in decreased market value.

Other fruit trees;-

All kinds of citrus, litchi, peach, apple, guava, jamun, date, palm, apricot, quince, pear, almond, plum, loquat, phalsa, and cashew require pollination by bees.

Cultivated field crops:-

Pigeon pea, lentils, clovers, Lucerne, mustard, rape, linseed, sesome, gingelly, buck-wheat, Cambodia, safflower, millet and sunflower require pollination by bees.

Vegetables:-

All cucurbitaceous plants, okra, beans, turnip, radish, onion, brinjal, and sweet potato require pollination by bees.

Timber tress:-

Neem, Cassia fistula, Acacia, Albizzia spp., Kachnar *(Bauhinia purpurea)*, eucalyptus, sandle-wood, raintree, wild cherry etc., require pollination by bees.

Natural and ornamental flowers:-

Cosmos, shoe flower, Golden rod, Cup and saucer, Tecoma stand, zinnia, coral creeper, rose, rangoon creeper, aster, wild rose, hydrangea, violet, portulaca, poinsettia, honey suckle, corn flower, coreopsis, dandelion etc., required pollination by bees.

Benefits of honey bee pollination in horticulture:-

- Bee pollination results in a higher number of fruits, berries or seeds. It may also give a better quality of product, and the efficient pollination of flowers may also serve to protect the crops against pests.
- Some types of crops have flowers that may only be pollinated during a short period. If such a crop is not pollinated during that time, the flowers will fall and no seeds, berries or fruit will develop. There have to be sufficient number of bees in the pollination crop.
- The value of bee pollination in Western Europe is estimated to be 30-50 times the value of honey and wax harvest in this region. In Africa, bee pollination is sometimes estimated to be 100 times the value of the honey harvest, depending on the type of crop.

Important Questions:

1. Write an essay on bees as pollinators.

2. Give an account of modern methods in employing honey bees for cross pollination.

3. Describe various qualities of honey bees that make them good pollinators.

4. Give an account of management of honey bees for pollination in horticulture garden.

5. Give an account of various effects of bee pollination on crops.

6. Describe salient natural and ornamental flowers pollinated by honey bees.

UNIT - 7

Products Of Beekeeping

Synopsis - (Honey; Bee wax; Bee venom; other products-royal jelly; Propalis. Steps of Honey extraction; Economic importance of honey; Bee wax; Production of Bee wax; Bee keeping industry; Recent efforts; Synthetic honey).

Introduction

Apiculture or bee keeping is the science of rearing honey bee for the commercial production of honey, wax, bee venom, royal jelly etc, as shown in a Fig. 7.1.

Apiary is the place where honey bees are cultured. It is also called as Bee yard.Bee and their pollination services contribute to maintaining biological balance in nature and enable various animal and plant species including humans, to thrive. They also provide bee products that are entirely natural food source. People have used them since time immemorial, and they are a particularly suitable source of food in today's increasingly faster pace of life.

Bee Product :- The following products from bees are obtained which are used by humans for different purposes.

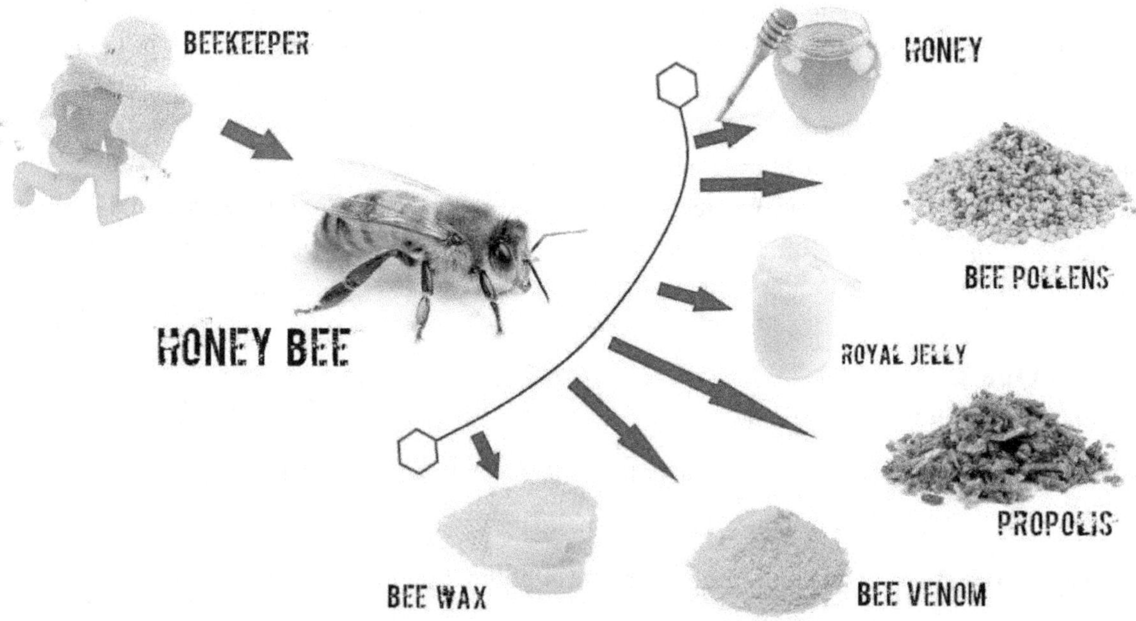

Fig. 7.1 Salient honey bee product

HONEY :-

Honey is a food material for the bees and their larvae. Large quantities of honey are stored in the hive to meet demands in scarcity. Chemically, honey is a viscous water solution of sugar.

Composition of honey and its different flavours depend on the kinds of flowers from which the nectar is collected. Nectar is sucked from flowers and mixed with saliva. It is swallowed into a special region of the gut, called honey stomach. Nectar is a disaccharide(sucrose). It is hydrolyzed by the salivary amylase to produce mono-saccharides (fructose and glucose). Therefore, honey is a sweet aromatic compound collected and synthesized by honey bees. In practical terms, honey is the digested, regurgitated and fanned nectar. Honey has got high viscosity and its density is high which is about 1.41. Honey can be crystallized or granulated. Honey having low glucose/water ratio will not crystallize. The colour of the honey varies depending on the original floral source.

Honey is sweet natural product of nector processed by honey bees. Honey contains the following natural components:

1. Levulose or Fructose (Ketose sugar) = 38%

2. Dextrose or glucose (Aldose sugar) = 31%

3. Surcose (Disaccharide) = 3.1%

4. Other sugars = 9%

5. Total acids like citric acid gluconic acid and HCl = 0.57%

6. Ash (Source of carbon) = 0.17%

7. Nitrogen = 0.04%

8. Water = 17.0%

9. Vitamins like Thiamine Riboflavin; Niacin; pantothenic acid; Vit. C; and Fat soluble vitamins (A, D, E and K) = Trace amount

10. Enzymes like invertase; diastase, glucose oxidase; Acid phosphorylase etc. = Trace amount

11. Minerals like copper; iron; phosphorus; silica; sodium, calcium, zinc etc. = Trace amount

12. Antioxidants like catalase, ascorbic acid and Flavonoids = Trace amount

Extracted honey in bowl.

Inside the hive the workers regurgitate the processed nector. The honey thus produced is still very dilute. After placing this honey onto the storage cells of the hive the bees "fan" with their wings to evaporate the excess water and bring the honey to its required concentration. Chemical composition of honey is illustrated above. To sum up, bees collect nectar from different angiospermic flowers, They swallow it into crop and mixes with the secretion of hypopharygeal glands, The secretion mix with the nectar. The enzymes digest the nector. The digested nectar is regurgitated into the storage cells. It is fanned by the workers bees to evaporate the water content. Therefore, natural honey is the digested, regurgitated and fanned nectar.

Bee wax:-

Beeswax is ideally made by young bees, precisely 2 to 3 weeks old. They prepare the beeswax after feeding the young brood on royal jelly, just before they venture outside the hive to forage. The beeswax is sourced from underneath the abdomen of the worker bees. It is used by the bees as building blocks of the hive. The wax is used in building comb cells for the young bees. It is also used for sealing cracks in the hive and building storage honey cells. Furthermore, bee wax acts as a cushion against infection for the brood.

Wax is secreted from wax glands situated in basal portion of the abdomen. It is pale yellow. It's specific gravity is 0.955-0.975. Some other properties of wax are 62-65° C melting point, 15-19 acid value and 70-80 ester value. Chemical composition of bee wax is as follow:

Chemical composition of bee wax-

Chemical Constituent	Per cent (Approx.)
Alkyl esters of wax and fatty acids	73.0
Cholesteryl esters of fatty acids	1.0
Lactones	0.5.
Free wax acids	13.0
Hydrocarbons	12.0
Moisture	1.0

Note: The wax from *Apis mellifera* is of good quality. In India, 70-80% of the wax is produced from *Apis dorsata.* The wax produced from *A. dorsata* and *A. indica* has low value but high ester value.

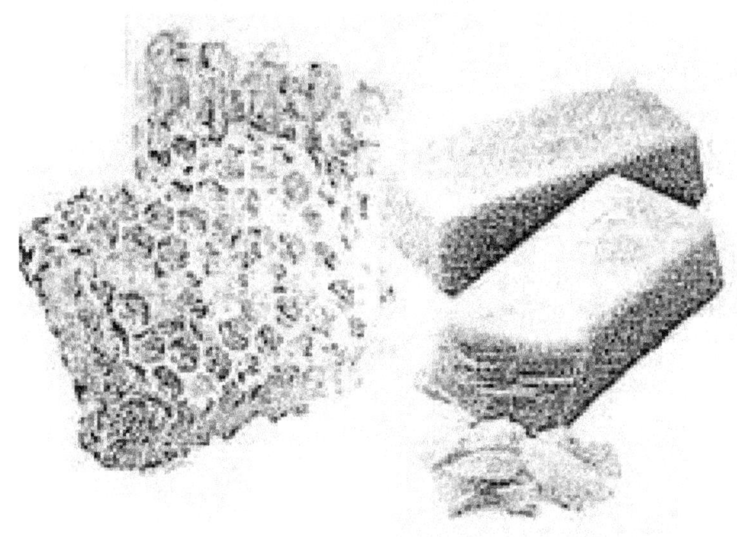

Fig 7.2 Bee wax pieces

Bee venom:-

Bee venom is a colorless, acidic liquid. Bees excrete it through their stingers into a target when they feel threatened. It contains both anti-inflammatory and inflammatory compounds, including enzymes, sugars, minerals and amino acids. In whole life, a bee can sting only sigle time because sting along with abdomen breaks.

Other Products of Beekeeping:-

The following are the other products of bee keeping:

1. **Royal Jelly**
2. **Propolis**
3. **Bee Bread**
4. **Royal jelly:-**

   Royal jelly is a milky colloidal substance and is produced by worker bee, especially for prospective queen larva. It is used as medicine for the treatment of human influenza, high blood pressure and respiratory infection. It is used as a component in skin care and natural beauty products. It is believed to have anti-aging properties and used regularly in popular alternative folklore medicine.

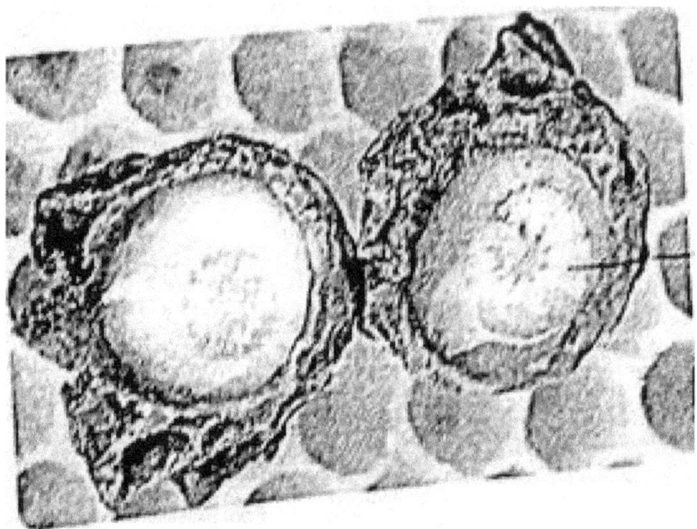

Fig. 7.3 Royal Jelly

1. Propolis:-

Honeybees repair their damaged comb by a wax substance, called propolis. It has medicinal importance for human as medicine for tonsillitis, bronchitis and dental infection are made by this substance.

Fig. 7.4 Propolis

2. Bee bread :-

Angiospermic plant pollens are collected by the field bees. As the bee is gathering nectar from flowers, the pollen from the stamen of the flower sticks on the hairs of the field bee.

These pollens (male gamete) are collected in the pollen sacs located on the legs of the bee. The house worker bees mix bee saliva, the plant pollen, and nectar. The bees secrete enzymes into the pollen. This mixture is known as bee bread.

Mechanism of Production of honey:-

Honey is a sweet, viscous food substance made by honey bees and some related insects, such as stingless bees. Honey gets its sweetness from the monosacchrides fructose and glucose, and has about the same relative sweetness as sucrose (table sugar). Fifteen mili liters of honey provides around 190 kilo Jules of food energy. It has chemical properties for baking and a distinctive flavor when used as a sweetener. Most microorganisms do not grow in honey, so sealed honey does not spoil, even after thousand of years.

Mature honeybees collect nectar from plants blossoms. Nectar is 80 to 95 percent water and 5 to 20 percent sucrose. As the bee transports the nectar back to hive, a conjugated enzyme in her honey stomach, called invertase, breaks the sucrose down into the two simple sugars, fructose and glucose.

Young bees remove water from the sugar solution using two methods. They pass the nectar from bee to bee and 'drink' the water out of the nectar by operating it through their stomach wall. They also create heat and air flow in the hive by vibrating their wings and flight muscles, thus evaporating water out of the nectar which has been stored in open cells.

When most of the sucroce has been converted to fructose and glucose and enough water has been dehydrated out of the mixture to bring it approximately 17.8% water content, we have a delicious sticky mixture, called honey. After honey is made, bees cap it with beeswax to maintain the low moisture content.

Varieties of Honey

The flavour, colour, texture, and aroma of honey depends on which plants the bees gasthered nectar from. Blueberry blossom, is dark amber in colour, and has a brown sugar after taste. Raspberry blossom honey is extra light in colour with a slight fruity taste, and buckwheat honey is almost black in colour with a heady, pungent odour and flavour. Generally, a dark honey will have a higher nutritional content than a light honey.

There are hundreds of flowering plants in the world that produce nectar, thus providing the potential to have hundreds of types of honey. Wild flower honeys represent a blend of flowering plants, and varies in colour and flavour depending on region and season.

Type of honey:- The following are the basic types of honey:

Comb honey:

It is untouched by human hands. It contains all the goodness that nature has put into the honey. It is bit awkward to chew.

Raw Honey:

It is extracted and cleaned honey using a settling tank at room temperature. It contains virtually all the goodness that nature put into the honey. It may granulate quickly and may separate in the jar with liquid fructose on top and granular glucose on the bottom.

Liquid Honey – (Filtered with minimal heat):-

It is extracted and cleaned honey using a 50 micron filter. Honey is heated to the same temperature inside a hive on a hot day. It contains a great deal of the goodness that nature put into honey. It may granulate in two to six months, depending on the type of flowers the bees visited to gather the honey.

Creamed Honey:-

Creamed honey is made from pure liquid honey through a controlled crystallization process to produce very fine uniform crystals, thus resulting in a creamy smooth consistency. Creamed honey has nothing added and has the same nutritional value as its liquid counterpart.

Liquid Pasteurized Honey:-

It is extracted and cleaned honey using flash heating to a high temperature, super filtered through a 1 to 5 micron filter, and quickly cooled. It loses much of the goodness that nature provided, but will last over 9 months on the store shelf without granulating.

Steps of Honey Extraction

Various steps for honey extraction are given below in precise form.

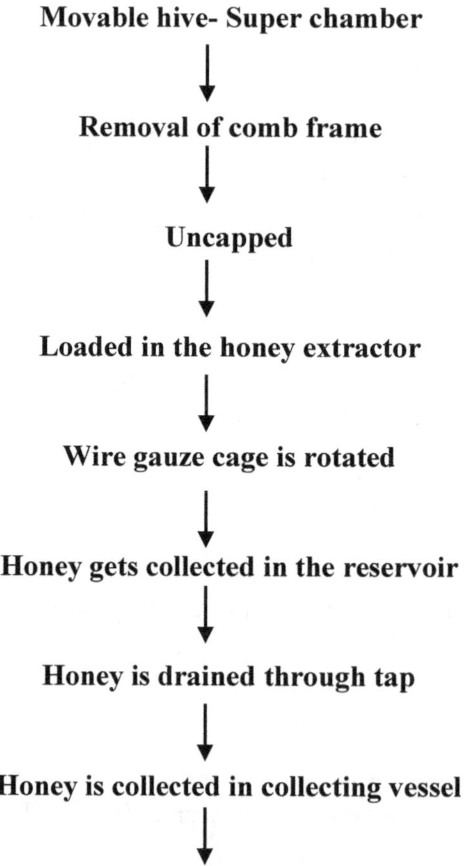

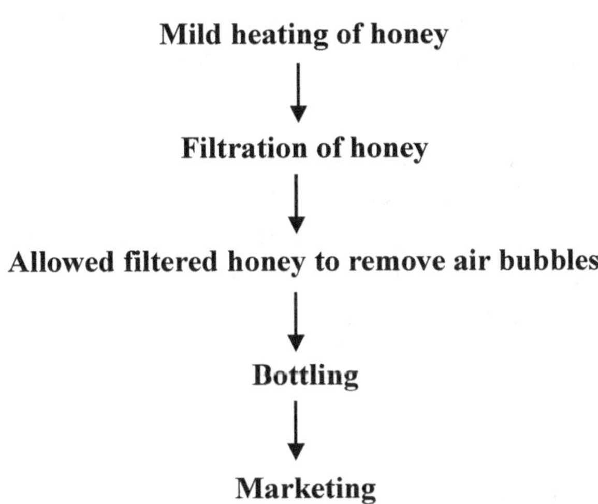

Fig.7.5 Steps showing honey extraction for marketing

Economic importance of honey

Honey is used by human beings as food and medicine.

1. **Food value:** 100 gm of honey provides as much nourishment as 6 liters of milk or 800 gm cream or 170 gm meat.

2.1 gm of honey provides 67 Kcal of energy. Sugar, minerals, vitamins and other vital elements of honey are easily absorbed by system. Honey can be taken by healthy as well as sick persons at any time in any season. It is used in the preparation of candies, cakes and bread. Honey is full of energy, since it contains much carbohydrates. It provides ready energy. It enhances appetite. Its natural sugars play an important role in preventing fatigue during exercise. Honey is useful in dryness of the mouth and body. Honey provides essential amino acids required for the growth of the body and improves metabolism. Minerals present in honey aid in blood circulation.

2. **Medicinal value:** Honey is mild laxative, antiseptic and sedative. It is generally used in Ayurvedic and Unani systems of medicines. Honey is quite helpful in building up of the haemoglobin of blood and also used as preventive against fever, cough and cold. It is also used to treat ulcers of tongue and alimentary canal. Germs of typhoid are killed by honey within 48 hours. Honey possesses anti-bacterial property. It can cure burns, cuts and sore throat. Honey acts as blood purifier. It also possesses anti – eosinophilic property. Honey prevents deposition of fats. It is good for kidney patients. Honey mixed with lime juice taken in the morning leans the fatty body.

BEEWAX

Beewax (cera alba) is a natural wax produced by honey bees of the genus *Apis*. The wax is formed into scales by eight wax-producing glands in the abdominal segments of worker bees, which discard it in or at the hive. The hive workers collect and use it to form cells for honey storage and larval and pupal protection within the beehive. Chemically, beeswax consists mainly of esters of fatty acids and various long – chain alcohol.

The wax is formed by worker bees, which secrete it from eight wax – producing mirror glands on the inner sides of the sternites (the veral shield or plate of each segment of the body) on abdominal segments 4-7. The sizes of these wax glands depend on the age of worker, and after many daily flights, these glands gradually begin to atrophy.

The new wax is initially glass-clear and colorless, becoming opaque after chewing and being contaminated with pollen by the hive worker bees, becoming progressively yellower or browner by incorporation of pollen oils and propolis. The wax scales are about three milimeters (0.12 inch) across and 0.1mm (0.0039 inch) thick, and about 1100 wax scales are needed to make a gram of wax. Workers bees use the beeswax to build honeycomb cells. For the wax-making bees to secrete wax, the ambient temperature in hive must be 33 to 36 °C.

Specificity of Bee Wax

In fact bee, wax is the eater of fatty acids. It is sticky and insoluble in water. Bees wax is created by worker bees to construct the comb. It is the 'bee sweat'

Worker bees have eight wax glands on the inner sides of the sternites (the ventral shield plate of each segment of the body) of abdominal segments 4 to 7.

The size of these wax glands depends on the age of the worker and after daily flights these glands gradually atrophy.

The ambient temperature for wax secretion in the hive has to be 33 to 36°C. Wax is synthesized by reducing sugar of nectar.

It is secreted in the form of wax scales.

The new wax scales are initially glass – clear and colourless, becoming opaque after by mastication the worker bee.

The wax scales are about 3mm across and 0.1mm thick.

It is estimated that about 1.5 million wax scales are needed to produce one kg of wax.

Amount of honey consumed in producing 1 kg of wax may vary from 6.66-8.80 kg.

Mechanism of production of Bees-Wax

The bees wax is secreted by wax glands, located in the worker bees.

A worker bee bears has 4 pairs of wax glands. They are also called mirror glands.

They are located one pair in each abdominal segment from 4 to 7.

The wax glands are situated just above the ventral plates (strenal plates) of the abdomen of worker bee.

In fact, the wax glands are modified epidermal cells.

Each wax gland is formed of three types of cells namely-

Epithelial cells; Oecocytels and Adipocytes

Each gland bears 30 to 50 pore canals. The pore canals pass through the body wall and open on the ventral side of abdomen by wax pores.

The ventral side of the sternal plate, situated below each gland, bear a polished cup like container called wax mirrors.

On this bases a worker bee bears 4 pairs of wax mirror from segments 4 to 7.

Fig. 7.6 Worker bee showing wax scales.

The sternal paltes of segments 4 to 7 are called mirror plates. Each mirror plate has a pair of wax mirrors.

Each mirror plate contains a front and a rear section. The front section is the wax mirror; which is smooth, shining and without bristles.

The rear section is formed of a bunch of bristles. That cover the wax mirrors of the following segment like the shingles on a roof.

The wax glands secrete wax in the form of a liquid. The wax liquid passes out through the canals and wax pores. It get collected in the wax mirrors.

The liquid hardens to form flakes called wax scales on the wax mirror.

The wax scales look like the dried flakes called wax scales on the wax mirror.

The wax scales look like the dried flakes of skin on a human head produced by dandruff.

The bees scrape wax scales with their legs, bring them up to their mandibles and chew them into wax.

The wax is applied for moulding the comb.

The production of wax by worker bees is shown in Fig. 7.7

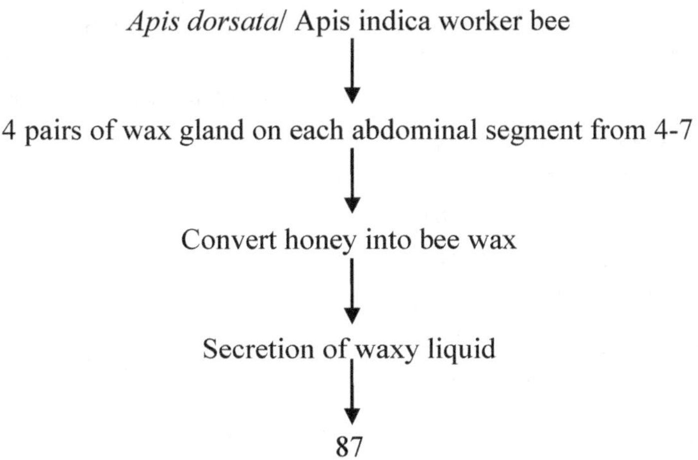

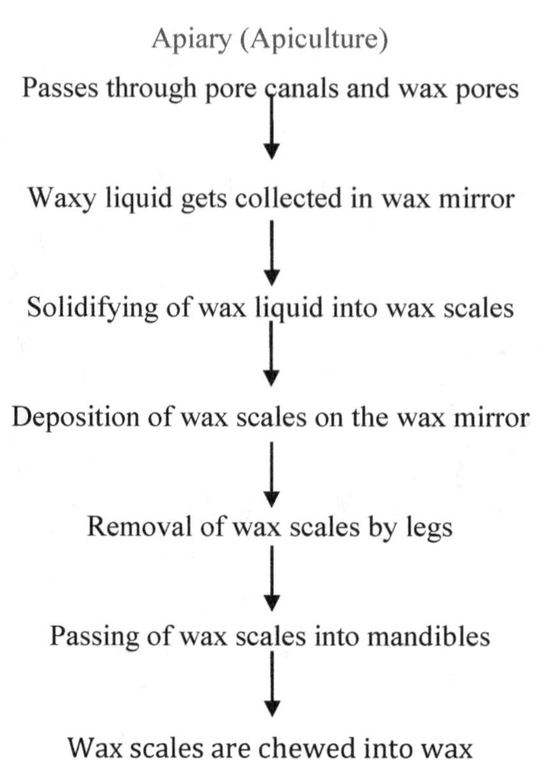

Fig. 7.7 Steps in the production of wax by worker bee

The wax glands synthesize wax from the sugar of honey.

7 kg of honey is used to produce 1 kg of wax.

One wax-scale weighs about 1 mg. About – 1000 scales are required ro make a gram of wax.

One million scales are needed to make one comb.

Wax glands are devloped in 12 to 18 days old workers. They are well developed in those workers who remain in the nest and construct combs. Wax glands degenerate in older workers who do not participate in nest building and maintenance.

Salient properties of Bees-Wax- The following are the salient properties of bee wax:

- Bee-wax is the ester of fatty acids and alcohols. It is the 'bee sweat'
- Fatty acids constitute about 70-74% of the total bee-wax.
- Other components found are free acids and hydrocarbons.
- Pure wax is white in colour and the yellow hues in the wax combs are caused by the soluble caroteniod pigments originated from the pollen.
- Bees wax is a tough wax formed from a mixture of several compounds including acid esters, acid polyesters, free acids and free alcohols.
- It also contains Vitamin A, which is a fat soluble vitamin.
- Bee wax is sticky in nature, without any taste
- The melting point of bee wax varies from 61-66°C.
- It's relative density at 15°C is 0.958 to $0.970/cm^3$.
- It's electrical resistance ranges from 5x10-20x10 ohm M.
- It's thermal conductivity co-efficient it 2.5x10Jcm/c.

- The sopanification value is 85-100.
- Bees wax is an inert material with high plasticity at relatively low temparature 32°C
- Bees wax is insoluble in water but is soluble in organic solvents such as benzene, benzol, chloroform, turpentine oil.
- Bee wax is resistant to several acids.

Methods of extraction of Bees-wax

Conversion of comb materials into the wax is called extraction of bees-wax.

Bees wax is extracted by the following methods:

1. Hot water extraction method

2. Steam wax extraction method

3. Wax press method

4. Wax rendering unit method

5. Submerged brood chamber method

6. Submerged sack method

7. Solar wax extraction method

1.   Hot water Extraction method

In this method pieces of old combs are washed and properly cleaned. They are placed in a drum having water. It is then heated to boil.

The comb materials start melting. The liquid wax floats at the top, which is filtered.

It is then cooled to make cakes.

2. Steam Wax Extraction method

In this method, the wax is melted in an insulated tank by steam blown directly into the mass of wax combs. The combs start melting.

The melted wax drops down and settles in the pan placed at the base of the extractor.

It is now cooled and shaped into cake.

3. Wax Press Method

In this method, pieces of combs are taken in a thick cloth bag.

The mouth of the bag is tied and the bag is placed in between two plates of press in a hot water drum.

During pressing, the liquid wax is filtered through the bag and the filtered liquid wax is collected and cooled to make cakes.

4. Wax Rendering Unit Method

The wax rendering unit comprises of a metal drum, heating elements, thermostat, a side funnel, a tray and grill pieces.

A filter cloth or wire mesh is tied on the top of the drum.

Water is heated for the liquification of wax.

Water is added from side funnel to the raise water level.

Liquid wax overflows through the outlet into the side grill which is cooled in the grill to become wax cake.

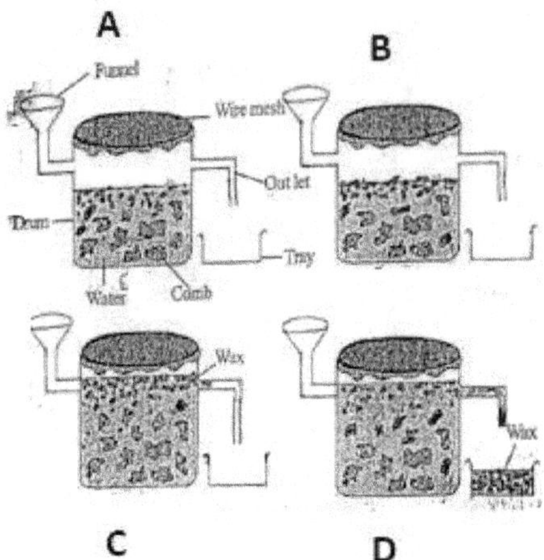

Fig 7.8: Wax rendering method.

5. Submerged Brood Chamber Method

In this method, the brood chamber of the hive is filled with comb. The top and bottom are closed with suitable wire screen.

It is submerged for several hours in tank of boiling water and the tank is then cooled.

The wax hardens.

The brood chamber is taken out.

The hardened wax is removed from the brood chamber.

6. Submerged Sack Method

In this method, the broken pieces of combs and stones are transferred in a sac. It is submerged in a drum of boiling water.

The sac is agitated well, so that the wax floats on the surface.

It is filtered and cooled to get wax pieces or cakes.

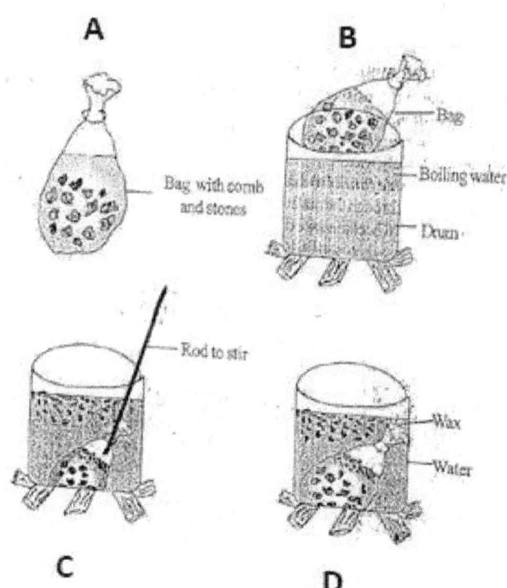

Fig 7.9: Submerged sack method.

7. Solar Wax Extraction method

In this method, the wax is extracted safely in a solar wax extractor. It is a type of wooden box. The top is covered with a glass plate. The box is having a galvanized metal plate.

The inside and outside panels are painted black to absorb the heat.

Comb materials are kept inside a container. The container is placed inside the extractor. Extractor is placed in open sun.

On a sunny day, the wax extractor may produce a temperature of 60° C which may melt the bee comb.

It proves to be the best method for the extraction of wax.

This has advantage of natural bleaching without contamination of any chemicals. There is no possibility of extraction of volatiles due to over heating.

Iron, copper or unprotected steel containers should not be used for processing of bees wax to prevent discolouration by reaction with these metals.

Salient uses of Bees-Wax: The following are the salient uses of bee wax:

- Oil obtained from bees wax is used in the treatment of wounds.
- Bees wax may be used to treat lupus.
- Bees wax may be used to prepare ointments, lotions, etc.
- Bees wax is rich in Vitamin A which is a fat soluble vitamin.
- Bee wax is an ingredient of shoe polish, sealing wax, glue, oil paints, pigments etc.
- Bees wax is used for preparing models for medical purposes.
- Bees wax is extensively used in sculptures.
- Candles are prepared from wax.
- Bee wax is used to prepare comb foundation sheets.

- It is also used to prepare water proof materials.
- The bees-wax is extensively used to coat papers and to strengthen threads.
- It is used to fill the pores in the wood.
- It is commonly used in making drawing crayons.
- Bee wax is highly used as an insulator in electronic goods.
- Bees wax has many industrial applications, like:

1. Foundry works

2. Engineering industries

3. Railway transport

4. Textile and leather industries

5. Perfumes

6. Confectionary etc.

Note: Bees wax (cera alba) is a natural wax produced by the honey bees of genus *Apis*.

BEE VENOM

Honey bee venom contains protein toxin secreted by the sting of honey bee.

Bee venom is also called apitoxin since bees are included in the family Apidae. Bee venom is produced by worker bee alone.

It is a colorless clear liquid.

It is sweet but with little bitter taste. The use of venom is to protect themselves from predators and enemies.

A newly emerged bee has very little venom. But the amount gradually increases with growing period.

The venom is produced by two glands namely poison gland and an alkaline gland associated with the sting.

Salient properties of venom: Honey bee venom possesses the following salient properties:

1. Transparent

2. Anticoagulant

3. Soluble in water

4. Insoluble in alcohol and ammonium sulphate

5. Bee venom contains bioactive compounds such as histamine, melttin, dopamine, formic acid and enzymes.

6. It exerts toxicity in victim tissue.

7. The dried venom is a transparent mass resembling gum Arabic.

8. It is extremely thermo-stable.

9. It retains toxicity for 5-6 days in the victim.

Composition of Bee Venom

The principal components of venom are- proteins, enzymes, peptides, amines, sugars, phospholipids, acids, minerals etc.

Proteins: Mellitin and Apamin are the specific proteins found in the bee venom. Mellitin is an antifungal, antibacterial and anticancerous compound.

Enzymes: Following are the salient enzymes found in the bee venom:

(i)   Phospholipase

(ii)   Hyaluronidase

(iii) Phosphatase

(iv)   α Glucosidase

Peptides: Following peptides are present in bee venom:

(i)   Mast cell degranulating peptide

(ii)   Scapin

(iii) Procapine

Amines: Following amines are present in the bee venom:

(i)   Histamine

(ii)   Dopamine

Sugars: Important sugars present in the bee venom are the following:

(i)   Glucose

(ii)   Fructose

Amino acids: Following amino acids are present in bee venom:

(i)   Aminobutyric acid

(ii)   α-amino acids

Acids: Important acids present in the bee venom are the following:

(i)   Formic acid

(ii)   Hydrochloric acid

(iii) Ortho phosphoric acid

Minerals: Important minerals present in the bee venom are the following:

(i)   Sulphur

(ii)   Magnesium

(iii) Calcium

(iv)   Copper

Methods of extraction of Bee Venom

Separation and collection of bee venom by squeezing a bee until a microdrop of liquid material is collected from the tip of the sting is called extraction of bee venom.

Method of extraction of bee venom is done in the following steps:

Different methods of extraction or collection result in varied compositions of the final extracted products.

Collection of venom is done under water to avoid evaporation of volatile compounds, present in the bee venom.

Bee venom is obtained from bee by giving low strenth electric shock.

The apparatus applied for venom collection is called cornel venom collector. It consists of a wooden frame, copper wire mesh, a thin plastic or rubber sheet and a glass plate.

The plastic sheet is placed above the glass plate. The copper wire mesh runs just above the plastic sheet.

The copper wire is connected to a 8-12 volts battery and then the apparatus is kept on the entrance of the hive.

When the bees come in contact the wire grid, they receive a mild electric shock.

Bees sting the surface of the plastic sheet as they observe this to be a source of danger.

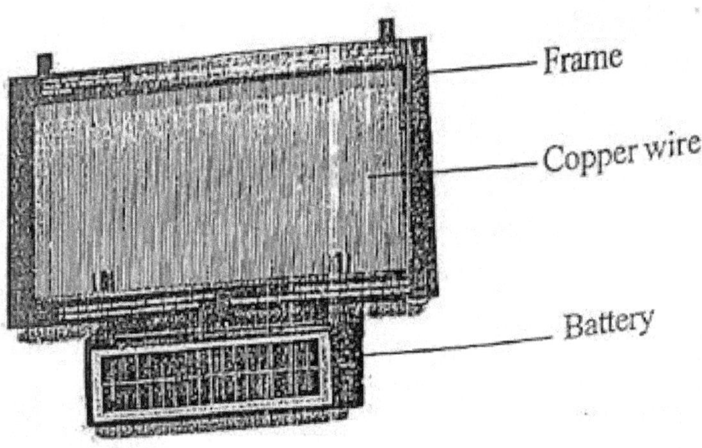

Fig 7.10- Corner venom collector

The venom is deposited between the glass and the protective material. Now the venom is dried and is later scraped off. It is to note that a honey bee stings single time in its life. After stinging, the abdomen is broken and bee dies.

Applications of Bee Venom: The following are the applications of bee venom:

- Bee venom is called a 'curative toxin'. It is excellent therapeutic agent.
- It is used to treat rheumatism. Is has beneficial effect on central nervous system which is impaired in the disease.
- Bee venom is extensively used to treat diseases of iritis and iridocyclitis.
- It is used to cure neuritis and neuralgia.

Bee venom is used in folk medicine to cure skin diseases like tuberculosis of skin (lupus) and eczema. Other uses of bee venom are the following:

- To remove scare tissue.
- To cure hypertension.
- To cure asthma.
- To cure premenstrual syndrome (PMS).
- Used in Bone healing.
- To cure hearing loss.
- To cure head ache.
- To make ointments.
- Bee venom is used in curing pains at joints.
- It is used in skin therapy.
- It is used to lower the B.P.
- It may be used as pain relief tablets.

Note:- It is to note that bee venom is most expensive of all the bee products. One kg of bee venom costs upto 70-80 lacks which is bought by various pharma medicinal company. It may be the best source of income.

Bee-Keeping Industry

Bee-keeping popularly called as apiary has gained a reasonable position as an industry in U.S.A., Canada and Australia but in tropical countries including India, it is not growing as per need of the people. Before 1953 attention to bee-keeping was paid only by the State Governments but in the same year, all India Khadi and Village Industries Commision, started to pay attention on this project and it was controlled by Union Government itself. Due to the functioning of the central organization, bee-keeping industry was firstly spread in South India and in some northern states of India also. Now-a- days bee-keeping industry is country wide and is a excellent source of cottage industry. Till date limited research and development work has been done in this field which indicates that 'Apiculture' can be raised to the status of a viable occupation in tropical climates provided appropriate scientific and developmental efforts are generated. Because the tropical countries are very rich in bee fauna, the scientific development of apiary may be helpful in agriculture and may contribute for providing nutrition and employment to rural population and may help in raising the economic status of the people in general and rural people in particular.

Recent Efforts in Apiary

An international conference on 'Apiculture in Tropical Climates', organized by the 'World Crops' in collaboration with International Bee Research Association (IBRA) was held at London in 1976. In this Conference it was recommended that the research and development of apiary should be taken on priority basis in the tropical regions. The Second International Conference on Apiculture in Tropical climates was organised by the Indian Council of Agricultural Research, New Delhi in collaboration with the Khadi and Village Industries Commission, Department of Science and Technology and Indian National Science Academy at New Delhi in 1980. In this Conference Eminent National and International scientists and experts realised the significance and scope of apiculture and suggested further research development as a device for raising the industrial status of apiculture in India and other tropical countries.

Synthetic Honey

The idea of preparing the synthetic honey came to Yaron and his friends of Israel in 2022, February. It is because honey bees are dying off due to infection of viruses. Synthetic honey preparation came in mind by considering the words of Albert Einstein who said that if bees disappeared then humanity can not survive for more than 4 years. It is because our angiospermic (flowering) flora of earth depend upon bees for pollination. About 75% agricultural crops are pollinated by bees. Keeping this in view, Yaron and Efrot extracted the specific protein from the stomach of the bees. They took nectar of any desired plant and mixed both in a honey machine and prepared synthetic honey. The name of this company located in Israel is BEE-IO. It is to note here that a single bee can make 1/12 spoon of honey in its entire life.

Important Questions:

1. Give an account of various products of apiculture.

2. Describe the method of extraction of honey and give its economic importance.

3. Describe the method of wax extraction and give its applications.

4. Describe method of venom extraction from honey bee and give its economic importance.

5. Write short notes on the following

    A. Bee keeping industry
    B. Recent efforts in bee keeping
    C. Honey
    D. Honey and its chemical composition
    E. Applications of honey
    F. Royal Jelly
    G. Bee Venom and its uses
    H. Bee Wax and its uses
    I. Synthetic honey

UNIT - 8

Social, Communication And Swarming Behaviour In Honey Bees

Synopsis - (Queen; Drones; Workers; Honey bee dance language; Swarming process; Types of swarming; Methods of Swarm control.

I- Social Behaviours

Introduction

Social behaviour vary even more widely than communication systems. Not only are there wide differences between species; there are differences within species too. We can recognize two different kinds of groupings. Animals—or, for that matter, plants—may occur in groups, because of their individual responses to particular factors of the environment. For example, butterflies attached to a field of flowers or the attraction of flies to garbage become grouped, because of their individual responses to a food source. Such groups are called aggregation. Other groups are formed because of the response of individuals to one another. These responses are called social responses, and the group is known as a social group.

Aggregations and social groups have the advantages of reproductive efficiency. Males and females both are usually present and their behaviours can be correlated. There are other advantages, having to do with protection of individuals and with efficient utilization of environmental resources. It has been known for some time that there may be physiological benefits in grouping. Even in very simple invertebrates, such as protozoans, it can be shown that groups of individuals are more resistant to such things as poisons or irradiation than are single individuals. In fact, this is true for vertebrates, such as fishes. These effects are not simply physical ones; they often involve the secretion of substances into the medium. Many aquatic organisms are known to "condition" the water in which they live, that is, to alter it chemically and make it more suitable for their life. A group of individuals may accomplish this more rapidly for a given volume than an individual organism. Group action against predators, of course, is an important aspect of group life. So much so, many situations, humans learn more rapidly in groups, where they can see and imitate one another. Actually, formation of social groups usually begins with parental care in animals.

Animal societies based primarily on structural specializations are examplified best by insects, while those based primarily on functional specializations, by vertebrates. Cooperation among animals usually involves some form of altruism. The (symbiotic). cooperative relationships between members of different species are usually reciprocal. Cooperation among members of the same species may also be based upon Kin selection. Kin selection is especially important in that it suggests a biological basis for altruism. In fact, cooperation is the most common social behaviour:

Definition of Social Behaviour

According to Tinbergen, (1952)—"Positive interactions between one individual of a species with the other is called social behaviour". The study branch of social behaviour is "Sociobiology", which may be defined as systematic study of the evolutionary basis of all forms of social behaviour in all kinds of animals.

Basis of Social Behaviour

The following points form the important basis of social behaviour:

1. Sexual attraction has been proposed as one of the basic forces bringing about social organization.

2. Next basis for social organization is the reduction of anxiety.

3. Another possible basis for congregation or grouping of animals is mutual stimulation.

4. Parental care.

5. Genetic constitution.

Factors Regulating Social Behaviour

Once animals are in contact, the structure and limitations of the social order are strongly influenced by a number of factors that are not directly social in character.

Sensory and motor capacities, learned and species specific behaviours, communication capabilities, nervous systems, and environment, all play a significant role in regulating social behaviour.

Illustrations of Social Behaviour in Honey been

Social behaviour though is widely spread in various groups of animals, however, the present chapter deals with the social behaviour in insects Honey bee examples.

Highly developed societies occur among termites, ants, bees, and wasps. Each member of such population is adapted structurally to carry out specific function in the society, All social insects build variously intricate nests, and these societies are stratified into structurally distinct castes.

In each of the four groups, different species form populations of different degree complexity. Many insects, however, live in association with each other, but they dependent on each other, therefore, they are not referred to be social. Further, certain are gregarious. They form a dense but temporary population viz. various kinds of in accumulate or gather together near source of light during night, particularly in rainy season prove good example of gregarious phenomenon.

Since this phenomenon is purely temporary, therefore, it can not be mentioned as a social behaviour. In certain groups of insects, division of labour and specialisation of individuals have become rigid and elaborate. These insects are called social insects.

A concise description of social behaviour in a social insect(Honey bee is described here:

Social Life (Behaviour) in Honey Bees

A second group of insects with very well defined castes, complex and integrated colonies are the honeybees. Chemical communication also plays a role in controlling the production"! different castes in bee hives. Queen bees secrete from glands in their jaws a pheromone called queen substance. As in termites, workers are always with the queen, grooming and feeding her. This constant exchange of food and pheromones is critical to the social integration of colonics of social insects. It has been given the name trophallaxis. In honey bees, a high level of queens substance in the hive inhibits workers from producing special quarters in the hive, in which queens develop. These queen cells are much larger than the cells, in which worker bare develop.

Each colony has 5,000-50,000 honey bees depending upon atmosphere and place. Colony is well organized showing division of labour by three castes namely, queen, males or drones and workers (Fig. 8.1). The cell types in hive is shown in Fig.8.2

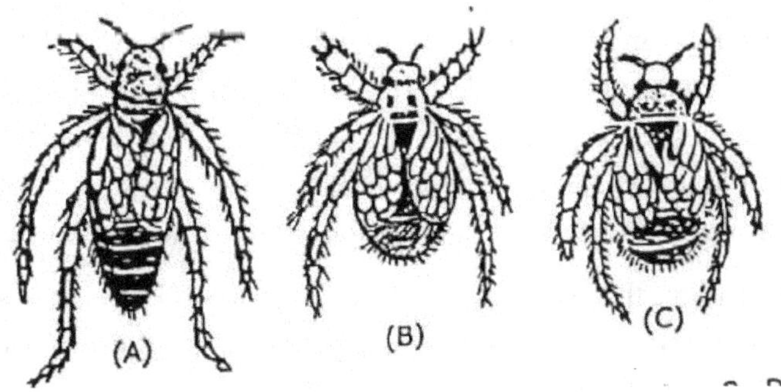

Fig.8.1 Different castes in Honey bee. A = Queen; B = Worker; C= Drone.

a. Queen

The queen is the only fertile female in the hive with immensely developed ovaries. Only one queen is found in each hive and feeds on Royal Jelly. Egg-laying is the sole function of the queen throughout her active life span. She is elongated, 15-20 mm long, and can be easily recognised by her long swollen tapering abdomen and short legs and wings. Structurally, she is incapable of producing wax or honey or gather pollen nector. The queen has an ovipositor on the tip of the abdomen, which is the egg-laying organ. A queen may survive for several years. It is said that the queen gets fertilized only once in her life, but in a single chance of fertilization, drone releases about two crore sperms, which are sufficient for the fertilization of the eggs at the time of lying by the queen throughout her life span. A research report from the U.S.A. reveals that out of 110 queens only 55 mated twice before egg-laying. It is also a fact that queen lays fertilized and unfertilized eggs, both in accordance to her will, but the factors governing such selective activity are still unknown. Queen copulates while it is on nuptial flight only in its life time. During nuptial flight only one male out of the drones copulates with the queen. Copulation completes in 15-20 minutes. During copulation the queen stores nearly 2 crore sperms in her spermatheca, which may live for 3 years. Queen returns to its colony after copulation. Here she starts laying eggs.

One queen lays about 1500-2000 eggs in a day depending upon the seasonal other ecofactors. The total weight of about 100 eggs is equal to her body weight. In the whole life span of two to five years, a queen lays about 15,00,000 eggs. When the queen in loses its egg-laying capacity, another worker of the same colony starts feeding on queen diet i.e. Royal Jelly and develops into a new queen and is provided with the facilities of real a At the same time old queen may be driven out. Sometimes some workers object that as to who the mother of the colony be driven out so ultimately they also come out with the mother Sometimes when two or three queens are developed in a colony, only one takes the place of the real queen and the others come out with some workers to establish new colonies.

It is rather important to mention here very clearly that in honeybees, the difference between queens and workers is nutritional. Workers are sterile females which have not had a diet permitting sexual development. Larvae in the larger queen cells are fed a special diet by the workers, including the substance known as Royal Jelly, which they secrete from special glands. On this diet they become queen bees. A newly emerged queen kills all the other larvae in queen cells. Therefore, there is normally only one queen bee in a hive.

b. Males or Drones

There are usually a few hundred (12-300) drones in a colony and also these are haploid fertile males which fertilize the queen and so called as KING of the colony. These are intermediate in size, 15-17 mm long, i.e. smaller than queen but larger than sterile females i.e. workers. These are considerably stouter and broader. Like the queen, they are fertile and their main function is to mate with the queen of their own or some other colony. Besides this function, they do nothing. They possess large eyes, small pointed mandibles and lack wax-producing glands, pollen collecting apparatus and a sting. Actually, they develop parthenogenetically from the unfertilized eggs. laid by the queen in a large Drone Cell.

Drones are totally dependent upon the workers and have been seen begging for honey from the workers. At the time of swarming, the drone follows the queen, copulates and dies after completing the copulation. The abdomen of the drones is deep black with lustre.

c. Workers

These are smallest bodied diploid sterile females. Although these are the smallest of the three castes but do all the works of the colony except laying eggs. Various adaptive features are well marked in the body of worker bees to perform variety of functions. These are well adapted to build the hive, ward off enemies, collect food, feed the queen and the drones and nurse the young. Actually, the workers are atrophied females and sacrifice even their whole life for the service of the colony. The total indoor and outdoor services or duties of the colony are performed by the workers only. That is why they are provided with some special adaptive structures for particular work:

1. Strong wings for fanning.

2. Long proboscis for sucking the nectar.

3. Pollen basket for the collection of pollen.

4. Ovipositor is modified into powerful sting to defend the colony against any attack.

5. Wax gland for wax secretion (Fig. 8.2).

6. Body hairs to adhere pollen grains.

7. Sucking type of mouth parts to gather water and nectar.

8. A large poison-storage-sac connected with the base of the sting.

9. Presence of a honey crop, a specialised part of the alimentary canal where nectar is chemically converted into honey with the help of the saliva.

10. Possess Nasonov Scent gland in their abdominal region. A pheromone is released from this gland which attracts other castes of bee. It also assists in marking the food and water sources by foraging individual and providing guidance to other bees to return back to correct home. It also assists to mark the nesting site, an essential factor for maintaining cohesion among bees.

Like the queen, they are also produced from the fertile eggs laid by the queen and live in a chamber called Worker Cell. It takes about three weeks in the development from the egg to the adult and the total life span of a worker is about one and half months i.e. about 6 weeks.

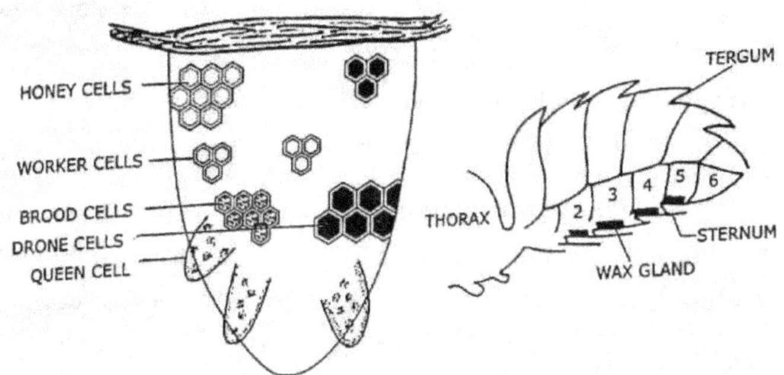

Fig. 8.2 A bee hive showing different type of cells and location of wax glands in a bee.

The workers which are engaged in outdoor duties, collect the nectar, pollen gum and water which are received and stored properly by the honey bees. The indoor workers are further sub grouped for specific duties. Some of them which are very sincere, attend the queen while some look after the nursery called as Nursery Bee. Some produce wax for the formation of the new hive, and are known as Builder Bee. The repairing of the comb is done by the Repairers. The dead body and other impurities are removed from the hive by the Cleaners. The fanning in the hive is performed by the Fanners. Several other functions like honey storage and ripening, are also done by the workers. The guard bee always watches at the gateway. It is said that upto half of life span, workers perform indoor duties, and later on become engaged in outdoor duties. It is to note that workers collect a resinous substance from the buds of the plants called propolis. It is used for sealing gaps between cells of the hive and repairing the hive A special type of propolis called balm is for polishing the internal surface of hive.

Formation of New Bee Colonies

When a hive becomes overcrowded, the queen honey bee together with some drones and many thousand workers recedes from the colony. The emigrants swarm out and settle temporarily on a tree or any other suitable place until a new hive is found. In the old hive meanwhile, the workers remaining behind raise a small batch of the old queen's eggs in large, especially built honey comb cells. These eggs develop into new queens. The first queen to emerge from its cell soon searches out the other queen cell and stings their occupants to death. If two new queens sometimes happen to emerge at the same time, they at once engage in mortal combat until one remains victorious. The young queen soon mates with one of the drones. In a nuptial flight high into the air, she receives millions of sperms, which are stored in a receptacle in her abdomen. The sperms from this single mating last through the entire egg-laying carrer of the queen.

Queen honey bee lays two types of eggs—fertilized ones with 32 chromosomes, giving rise to queen or workers, and unfertilized ones with 16 chromosomes producing drones by parthenogenesis. Eggs are long, cylindrical and light yellow coloured. Only one egg is laid down in every brood cell of the hive, which sticks to the wall of the cell. Hatching occurs after an incubation period of 3 days and larva emerges out of it.

Larvae emerging from all types of eggs are fed upon royal jelly by workers for first three days. Royal jelly is thin white secretion from lateral pharyngeal glands of young workers. Royal jelly consists of 45.15% protein, 35.55% fat, 20.39% sugars mainly glucose and levulose, 2-3% minerals and vitamins (all

B group vitamins especially pantothenic acid). Royal jelly is highly nutritive. For first three days, all larvae are fed on royal jelly, thereafter, the prospective queen larvae are further fed on royal jelly for next five days, but the larvae to be developed as drones and workers are fed on pollen mixed controlled diet for next 6-7 days. There queen, workers and drone brood cells are sealed on 8th, 9th and 10th day respectively by workers. Male brood cell seal is protruded upwards whereas seals of other brood cells are flat. Larvae develop into pupa in these sealed brood cells. After two week of pupal stages, young honey bees emerge out of cells. Thus, complete metamorphosis occurs in honey bees.

When a colony is overcrowded by honey bees, queen flies with some male and workers and forms new hive at another place and they start living there to develop a new colony. This process is called swarming. It occurs late in the spring or early summer when food is in abundance. It occurs normally in the morning. When queen bee leaves the hive, workers start feeding the growing bees on royal jelly so as to allow them to develop as queen. The first queen after her emergence from the queen brood cell kills rest of the queens still sealed in queen brood cells. The new queen starts nuptial flight with some male bees.

When the queen of the colony ceases to lay eggs or she dies, her position is taken by a new strong bee. This process is called supercedure.

In adverse conditions, all the members of the colony leave the place and reach another place. This process is called absconding. Before leaving the hive, egg laying is stopped, and the stored honey is sucked.

Paning winter when temperature is low, the worker bees densely cover the hive and rubthar legs to produce heat so as to maintain the temperature of the hive. In summer, when the temperature is high, the worker bees spray water drops over the live and start fanning the inne with their wings, so as to establish the temperature down by the process of evaporation. In this way, bees maintain the temperature of the colony by air conditioning,

When there is scarcity of food in the vicinity of the colony, its members attack other colonies and forcefully take out honey from them. Such phenomenon is called robbing.

II - Communication in Honey bees

Communication can be defined as influencing the activity of one individual by some behaviour of another or animal communication is the passage of information between two individuals. The animal which sends signal is called signaller and the animal that receives signal is called receiver. Animal communication may also be called as biological communication of those actions and activities of an individual that influence the behaviour of a second individual. Such a definition is too broad, because such activities as predation and parasitism also influence the behaviour of second individual. In general, the term communication implies something exchanged between an individual and its environment, both social and non-social communication is essentially a social phenomenon in animal societies. Allman (1962) defined social communication as a "process by which the behaviour of an animal affects the behaviour of others." Broadly speaking, in a communication system there is the sender and the other is the receiver where both are mutually benefited.

Situations in which communication among species is mutually beneficial usually involve Symbiosis. In one form of symbiosis, known as Commensalism, one species benefits from the relationship while the other remains unaffected. For example, the trumpet'lish (Aulostomus) sometimes joins schools of yellow sturgeon and takes advantage of this camouflage to approach smaller prey fish. It darts out from among the sturgeon and seizes its prey. The sturgeon remains unaffected by the association, and there appears to be no communication between the two species.

In true symbiosis or mutualism, both species benefit and communication usually occurs between them. For example, the honey badger (Mellivora capensis) lives in symbiosis with a small bird called the honey guide (Indicator indicator). When the bird discovers a hive of wild bees, it searches for a badger and guides it to the hive by means of a special display. Protected by its thick skin, the badger opens the hive with its large claws and feeds on the honey combs. The birds feed upon wax and bee larvac, to which it could not gain access unaided. If the honey guide can not find a badger, it may attempt to attract people. The natives understand the bird's behaviour and follow it to the hive. It is an unwritten law that the bird be allowed to take the beelarvac.

Communication between members of a single species is an important aspect of ethology i.e. animal behaviour. Animals ranging from insects to mammals are well known to pass or convey information to other members of their group about location of food, territorility, danger of predators, notification of rank, sex of the individual and readiness for mating. They do so by means of specific signals evolved for the purpose. The signals may be in the form of conventional signs, gestures, touch, sounds, light, "chemicals or a combination of these. Human also uses these method of communication, but he adds his own, a unique type of verbal language:06's speech. Therefore, human is on top in the list of communication patterns.

Essential Components of Communication System

Ewsbury (1978) reported that following seven components are essential in a communication system:

1. Sender: An animal that sends a signal. It may also be called "releaser." .

2. Receiver: An animal that receives the signals sent by senders as a result there is change in the behaviour of receiver.

3. Message: Signal like intention depicted by the sender of specific nature which is extracted by its receiver.

4. Channels: Represent pathways through which a signal normally travels. These are vocal auditory channels where different sensory receptors are involved. Important channels are: odour, sound, surface vibration, electric field, vision, touch, etc.

5. Contact: The setup under which a signal is released and received.

6. Noise: It is a background activity in the channel which is not related to the signal..

7. Code: A possible complete set of signals and contacts may be termed code.

Types of Signals in Communication System

There are two basic signals-1. Discrete (digital), and 2. Graded (analog). The alarm calls of various species, which are typically emitted with the same intensity each time and are relatively constant across species, permit communication among different species. These discrete signals of fixed frequencies and duration that makes them difficult for predators to translate Marler, 1957).

Graded signals may vary in intensity as a function of the strength of the stimulus, It seems important to mention that signals are not absolutely arbitrary.

Principles Involved in Communication

Communication usually involves two animals first, the signaller who sends out the message, and second, the receiver who receives and reacts. Both are complements to each other. Whatever, may be in the form of a message, the receiver is able to pick out the signals of its own species, usually from other massages that surround it. To receive a signal the recipient must have sense organs matched to the signal. .

The influence of the signal on the receiver's behaviour depends upon its own internal state, It also depends upon the context-the circumstances in which the receiver finds itself and not the signal alone. ..

Signalling that evokes mating behaviour as an example, may be simple enough to be stimulated by inanimate objects. The fruitfly (Drosophila subobscure) courts and attempts to mount moving wax balls of various sizes; for other species (e.g. D. ambigue and D. triahs) the movement of an inanimate object must be a particular form different from that for D. subobscure. Often releaser-evoked behaviour combinations are characterized by an interaction composed of a series of active signals and responses. The analysis of probable releasers involved in the mating behaviour of cockroaches (Brysotriafumigata) is an example of a case of interaction (Fig. 8.3) -

The particular releasers involved in this interaction are not as important as the fact that the mating pattern is the product of a series of interactions. The failure of either partner to emit the proper response or to respond to the signal of the other at one or more points in the process would make consummation of the union very improbable. Thus, a complex of behaviour governs a function vital to continuation of a species.

Communication behaviour, particularly those involved in the reproductive cycle; can also be the basis for the beginning of a new species.

Communication Systems

Communication is the primary link between individuals that binds them into an organized unit. The need for communication increases as social organization becomes more complex Without ability to convey information from one individual to another, social behaviour could not exist.

There are numerous communication systems that depend on touch, sight (vision), hearing, vocalization, chemicals, languages and even electrical fields. Various aspects in animal communication are here being described:

1. Chemical Communication.

2. Visual Communication.

3. Auditory Communication.

4. Tactile Communication.

5. Electrical Communication.

6. Language Communication.

7. Complex Communication Systems in Social Animals (dance language in honey bee).

8. Surface Vibration.

Dance language as a complex communication system in honey bee has been described.

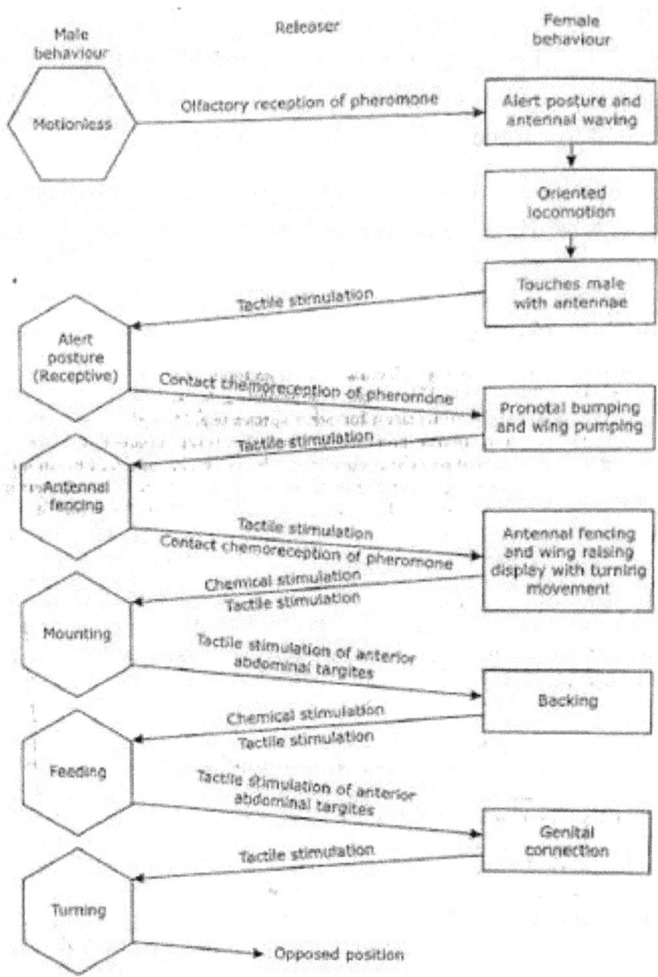

Fig. 8.3. The role of a sequence of releasers in the mating behaviour of the typical visecl).

Complex Communication System in Social Animal- Honey Bee (Dance Language)

Important as chemical communication is in the integration of colonies of social insects: perhaps the most interesting communication system, used by these organisms is the dance:language of the honey bees. Actually, bees have developed a complex language that utilize chemoreception. as well as tactile, visual and auditory stimuli. The operation of this communication system has been worked out in detail by the German ethologist K. von Frisch in 1967. He was honoured with Nobel Prize for this work in the year 1973, K.V. Frisch observed the behaviour of the bees inside a specially constructed hive with glass wall and noticed that returning foragers performed dances that attracted the attention of other bees. He identified two types of dariče a round dance (Fig. 8.4) and a waggle dance (Fig. 8.5), Both, types of dances are performed in the dark hive, so that the kind of food is determined by chemoreceptors on the antennae of the other bees, indentifying the pollen or nector on the dancing bee, and the location by placing the dancing bee and following the pattern of the dance. Sometimes a low humming noise (sound) is also associated with the dance, apparently also indicating distance. After repeating the dance a few times the bee then returns to the food source for more nector. As long as the supply remains large, the dance is repeated after each trip, but when the supply is limited the bce continues to make trips until the supply is exhausted, but does not perform the dance in the hive.

Round Dance: The round dance or rundtanz is used when the food is nearby. It consists of small circles, first to one direction and then the other as shown in fig. 8.4 after a bee repeats the dance a few times it leaves the hive and others, excited by the dance, follow they fly in circles around the hive until the food is locted.

Fig. 8.4. Round dance language of honey bee when there is a close food supply found

Waggle Dance: The waggle dance also known as "Schwanzeltanz" or Tail wagging dance takes the form of a flattened figure eight (Fig. 8.5). This dance occurs when the food source is further away than about 160-330 feet (average roughly 250). This dance informs the other bees of both direction and distance. It is preformed on a vertical surface in the hive and consists of three phases:

1. Semi circular walk.

2. Straight rush.

3. Final semicircular walk completing the circle. Bo?

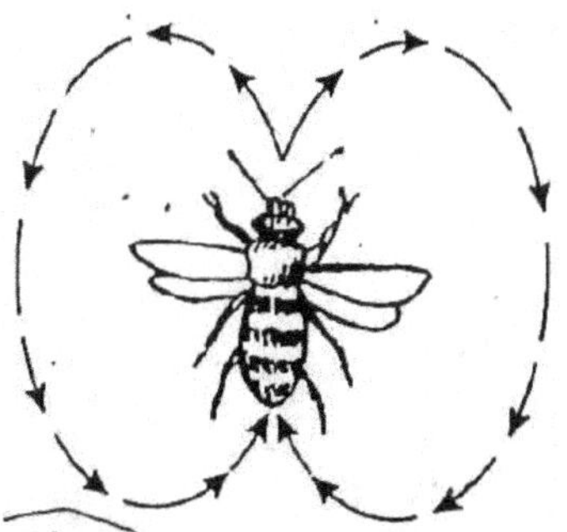

Fig. 8.5. The Waggle dance,

Direction of the food from the hive is indicated by the straight rush-up the side of the hive indicates towards the sun, down is away from the sun and diagonals relate to angles toward or away from the sun. The rate at which the dance is performed indicates the distance the hive, with slow rhythms of about four dances/minute indicating distances about 10 km (6 miles) away and 16 dances. Per minute about 1 km (016 mite) distance.

If the nectar is 90° to the left of the sun, the straight-line dance is horizontal with the bee running to the left. Actually, the plane of the waggle dance is simple, but it is rather difficult to describe in words. However, the following inferences in nut shell may also drawn from the waggle dance made by honey bees:

1.   Type of Food Available.

2.   Distance of the Food Source from the Hive.

3.   Distance of the Food Source in Relation to the Position of the Sun.

4.   The Amount of Food Supply Available at the Food Source.

1. Type of Food Available

When a forager or scout sits down on the food source, the scent of the food sticks to its body". Later on when the forager or scout performs its dance the other bees all the time soon its body with their antennae, consequently, they become aware with the scene of food sticked to the forager's body and know that what kind of food the scout has discovered.

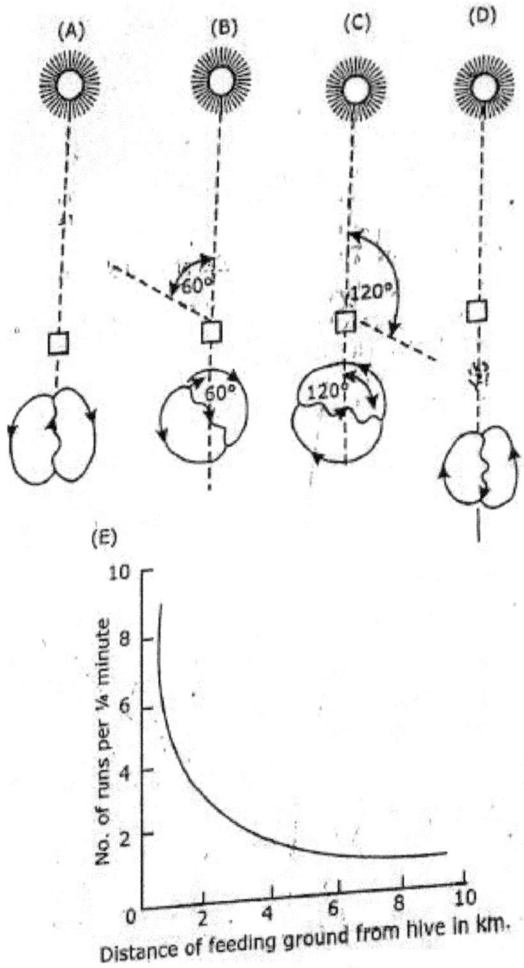

Fig. 8.6. *Use of Waggle dance (A), (B), (C) and (D), the position of a food source with respect to the sun is indicated by the angle of the straight portion of the dance. (E) The speed of the dance Indicates the distance of the food source from the hive (a diagrammatic representation).*

2. Distance of the Food Source from the Hive

If the food source is within about 160-330 feet from the hive then the bee performs dance, as a result, the other bees know that the new food source is present the hive.

When the food source is more than the aforesaid figure then the bee performs a waggle dance (i.e. to move side to side) making a figure of 8. As to how the actual distance is calculated by the scout bee and communicated to other bees is not well known. However, frequency of the waggle dance may give an idea of the exact distance of the food source. Actually, the frequency of the waggle dance is found to be inversely proportional to the distance between the hive and the food source. Evans pointed out certain figures by calculating the frequency of waggle dance.

If the food source is 100 meter away from the hive then the bee makes 60 waggle dance) per minute.

If the food source is 2000 meter away from the hive then the bee makes 13 waggle dance/ minute, and if it is 4000 meter then waggle dance reduces to 9/minute.

The sound pulses produced by the wings movement during the wagging dance also seems to have some relation with the distance. The number of sound pulses produced per waggle dance appears to be directly proportional to the distance between hive and food. It is believed that sensory organs present on the legs and antennae are capable of receiving these sound pulses. Evans has given some figures in connection with the pulses and the distance if it is 106 meter then the pulses are about 5.4 to 10.2 and in case the distance is 700 meters the sound pulse arc 38.8 to 43.6.

3. Distance of the Food, Source in Relation to the Position of the Sun

As shown in Fig. 8.6, the waggle dance also communicates the direction of the food source in relation to the position of the sun. When the food source is towards the sun, the waggle dance is directed upward Fig, 8.6A and when the food source is away from the sun the waggle dance is downward Fig. 8.6 D.

When the food source is situated on the straight line joining the hive and the sun then the waggle dance is directed upward and downward but when the food source is located at an angle, in that case waggle dance is made at the angle of the food source present in relation to the hive and the sun (Fig. 8,6B and C).

When the food source is at an angle to the right of the sun-hive line then the waggle run is at the same angle to the right of the line. If the food source is at an angle to the left of sun-hive line, then the waggle run is at the same angle to the left side of the line.

4. The Amount of Food Supply Available at the Food Source

The forager gives specific information to the colony about the rich or poor source of supply at the new food source. If the new source is very rich in food then the honey bee dances vigorously for a fairly long time but, if the food supply is poor then the dance is weaker and lasts for short period.

The dance of honey bee is a subject of extensive research, because much, is still unknown about this behaviour of the honey bee. However, the effectiveness of the dance is obvious beyond any doubt because of receiving the information through the dancing bee, the other bees reach the correct feeding place with remarkable exactness and surety.

Further Fig. 8.7 Shows an experiment wherein Von Frisch placed the hive on the side of a high ridge and the food on the other side. The bees did not fly over the ridge but around the end. The dance indicated the true direction of the food, but the distance was that of the trip around the end of the ridge. Bees following the dance from the hive in the direction indicated and encountered the ridge. They then flew around the end of the ridge, and onward the proper distance, encountering the food.

The communication by dance in bees is called a "Language because the dance is delayed transmission of the information to other bees, of a response to the environment, instead of the more usually the immediate response, such as cry of fright or a courtship song.

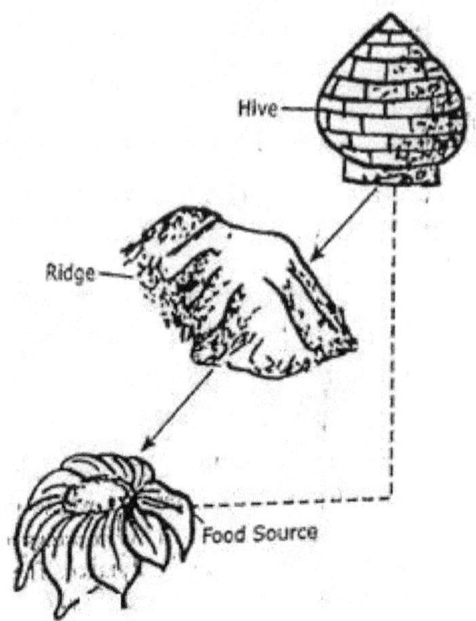

Fig. 8.7 Bees are able to communicate complex Ideas concerning the location of food, as shown by the experiment conducted by von Frisch.

Flower Recognition by Honey Bees

K. von Frisch observed that honey bees have well developed colour vision that differs from human vision in being insensitive to red, but that extends into ultra violet, where human eyes are totally insensitive. von Frisch discovered that many flowers have well developed markings called honey guides. Some of these markings are visible only in ultra violet light (Fig. 8.8).so they are normally invisible to human but visible to bees.

Experiments with honey guides indicate that small details of flower pattern can influence the behaviour of the honey bees (Manning, 1956).

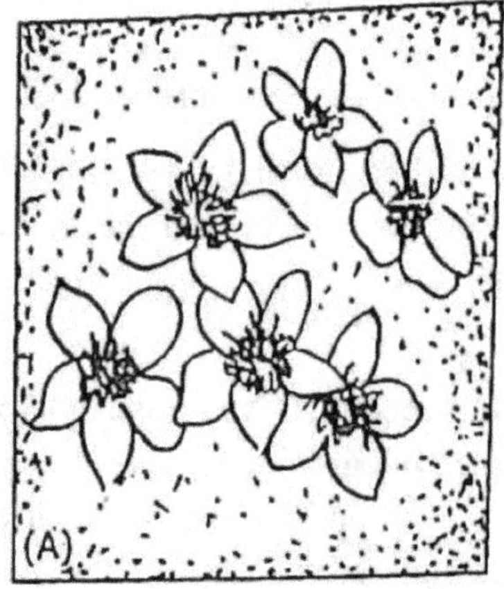

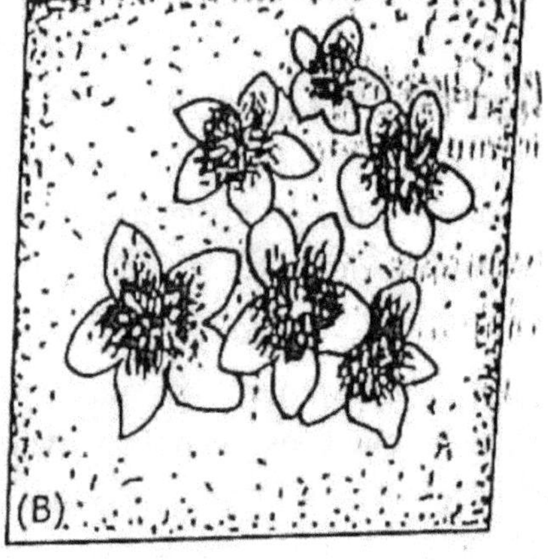

Fig. 8:8. *Flowers that appear to us to be white in day light (A) may have distinct patterns called honey guides when viewed in ultraviolet light (B), Bees are sensitive to ultraviolet light and respond to the honey guides*

Functions of Communication in General

Communication systems function in one of there ways: Primers, Releasers, and Stimuli for learners.

In the case of primers, the message acts as a stimulus that leads to internal changes, so that the organism is ready to act when further stimulated. A message can make the receiver more likely to act on or detect a subsequent message.

A communication system may also act as a "releaser". First and foremost, the concept of the releaser has been widely used by ethologists, Konard Lorenz and Niko Tinbergen. Releaser is considered to be a signal generated by one member of a species originating from an external source. Ordinarily, it evokes a response that is typical of the species for a given set of conditions. For example, the red breast of the robin is the releaser that leads to aggressive territorial behaviour by a male robin. Signals and responses in these cases may have very definite patterns and are called "ritualized" bahaviour. Releasers usually involve behaviour related to some survival or maintenance factor.

One other function of communication systems is to furnish stimuli, to which responses or classes of responses can be learned.

III Swarming Behaviour of Honey Bee

Detail study on swarming behaviour in honey bee has been made by N. Arumugam and his colleagues in his book "Applied Zoology". Their salient illustrations have been cited in the present section.

The mass flight of a part of honey bees from a large colony to start a new colony is known as swarming or the process of leaving of the colony is known as swarming.

The mass of bees making a swarming is known as swarm. The first swarm is led by the old queen but the second swarm is led by the 7-8 days old virgin queen which is followed by the male members (drones) and is called nuptial or marriage flight. One of the drones begins copulating with the queen in open air at a height of about 100-200 meters and fertilizes the queen and dies during the course of copulation. The queen receives spermatophores from the drones and stores them in the spermatheca. Along with the queen, died drone falls on the ground and the queen reaches the hive.

Swarming happens towards the end of spring or early summers but the real cause of swarming is still unknown. In summer season, when plenty of food is available and the hive is overcrowded by the bees the queen absconds the hive on a fine fore-noon with some old drones and workers and establishes a new colony at some other distant place. Now in the old hive, a worker is provided Royal Jelly and is, metamorphosed into a new queen of the colony. This new empress of the colony neighbor tolerates her successor, as a natural law in the hive. So she orders to kill the other sisters, if any, in the hive. When the egg laying capacity of the old queen is lost or it suddenly dies a new young and vigorous queen takes the position of the old queen and is known as supersedure.

A swarm may contain queen, worker and drones in thousands of number without a comb.

The principal function of a swarm is to protect the queen until scout bees can find a permanent location suitable for a new hive.

It is the natural means of reproduction of honey bee colony

Swarming produces two colonies, a parent colony and a daughter colony. The tendency to swarm varies from species to species.

Swarming is the natural innate or instinct behavior of the bee colony.

The instinct to swarm is very low in case of *A. mellifera* as compared to *A indica*.

The original colony may produce one or two or more colonies by swarming. Obviously a colony is weakened by swarming.

Swarming Process of a honey bee colony comprises of 30,000 to 60,000 bees.

All the bees in a colony are organized and controlled by the queen pheromone (a semio-chemical)

Trigger of swarming

Following are the triggers of swarming:

* Over crowding

* Less queen pheromone secretion

* Poor pheromone distribution

When the queen substance is in adequate amount, it can reach all the bees in the hive. It makes the entire colony to feel that they have a queen. Queen pheromone suppresses queen cell raising and swarming preparation. When the queen becomes old, her pheromone can not approach to all the bees. This condition also arises when there is overcrowding in the hive.

Swarming Preparations: Following preparations are made for swarming-

* Queen cups constructed

* Less foraging of workers

* Resting of workers

* Gorging honey

*Less brood rearing

* Queen fed less

* Queen fed on less protein diet

* Starving of queen

* Ovaries of queen lighten

*Light weight enables flight

When there is insufficient or less amount of queen pheromone, the workers feel that they are queenless and start building queen cells (queen cups).

This is the initiating step in swarming. The nurse bees start constructing queen cups.

The queen lays eggs in the newly constructed queen cups.

The queen cup is now called queen cell. As this queen cell is constructed for the purpose of swarming, hence is called swarm cell.

The egg hatches into a larva on the 4th day.

The nurse bees feed the larva with royal jelly.

The larva metamorphoses into pupa on the 8th day and the swarm cell is capped.

When the queen cells are capped, the mother queen and the worker bees start to leave the colony for swarming.

The nurse bees feed the queen a diet less in protein-This causes the queen to stop the egg laying. The workers store sufficient nectar in their crop.

Before leaving the old colony, a small group of experienced forager bees called nest-site scout bees find a nearby location for intermediate stop.

Swarming is initiated by a special type of dance, called the Schwirrlauf dance, or whir dance.

During the Schwirrlauf dance, the workers move without stopping in straight lines across the comb: -

They vibrate their partially spread wings.

These dancers also make occasional five second contacts with other worker bees.

They produce a "piping-signal" that primes the workers for swarming.

This makes the bees warm up their flight muscles for departure.

They produce a "buzz-run" signal which initiates the departure of the swarm.

60 minutes before swarm departure, piping signals gradually increases and reaches peak at the start of the departure.

The swarm comes out and aggregate in an intermediate stop which may be near the hive, tree branch, building corners, broken walls etc.

At intermediate stops, they cling together in a dangling clump.

In these temporary perches they have no protection from predators and bad weather, hence they search a new permanent place to settle soon.

In order to find a new location a quorum of scout bees (25-50 bees) are sent out from the temporary swarm.

These scout bees find possible nesting places for their permanent settlement.

The scout bees return and report to the other scout bees the location by representing a wagtail dance on the surface of the swarm.

The quorum of scout bees finds many places for settlement, but final settlement place is decided on the basis of group decision making.

Scout bees decide among themselves to finalise the best place.

When one site has been finally agreed on, the swarm flies to the new site, guided by scenting bees.

When the swarm of honey bees reaches at their new home, the workers start performing a ritual known as fanning,

During fanning, the workers raise their abdomen into the air and waggle it. By raising their abdomen into the air, the workers expose a white-tip called a Nasanov gland located at the hind and of the abdomen.

It releases a pheromone called "settling" or "orienting" pheromone into air which (pheromone) enables the bees to keep together to form a cluster.

The other workers detect the pheromone and interpret this as a message saying "This is our new home; hurry up and move in". So the process of fanning speeds up the move of other workers into their new hive.

The worker bees start constructing a comb, as their permanent venue.

In the queenless parent colony, the virgin queen emerges out. She goes in search of other queen and fights to death. She stings to kill the occupants of other queen cells. She signals her presence by making a sound called piping.

It makes the sound by pressing itself against the comb and vibrating its flight muscles without moving ber wings.

The successful virgin queen goes for nuptial flight on the 22nd day. She mates with drones and after this she returns to the parent hive and heads the colony.

She starts egg laying on the 27th day.

If the parent colony is too strong, the daughter queen may also vacant the colony with some workers as an after swarm.

A second daughter queen then takes over the colony in the same way as the first performed.

The swarm which issues from the parent colony is the "primary swarm". The subsequent swarms are known as "after swarms" or casts.

Issue of after swarm depends upon the strength of the colony, the availability of virgin queen etc.

There may be three to four swarmings from a colony depending upon its strength.

If the colony is weak after the issue of the primary swarm, the emerging queen destroys all the brood cells or kills the brood and remain in the parent colony and the activities of the are renewed. Process of swarming has been precisely illustrated in the Fig 8.9.

Incitation of swarming due to overcrowding and less queen pheromone secretion

↓

Poor distribution of queen pheromone secretion

↓

Some worker bees do not receive queen pheromones

↓

They construct queen cups

↓

Queen lays eggs in the queen cups

↓

Queen cell

↓

Egg hatches into larva on the 4th day

↓

Nurse bees feed the larvae with royal jelly

↓

Larvae metamorphoses into pupa on the 8th day

↓

The queen cell is capped

↓

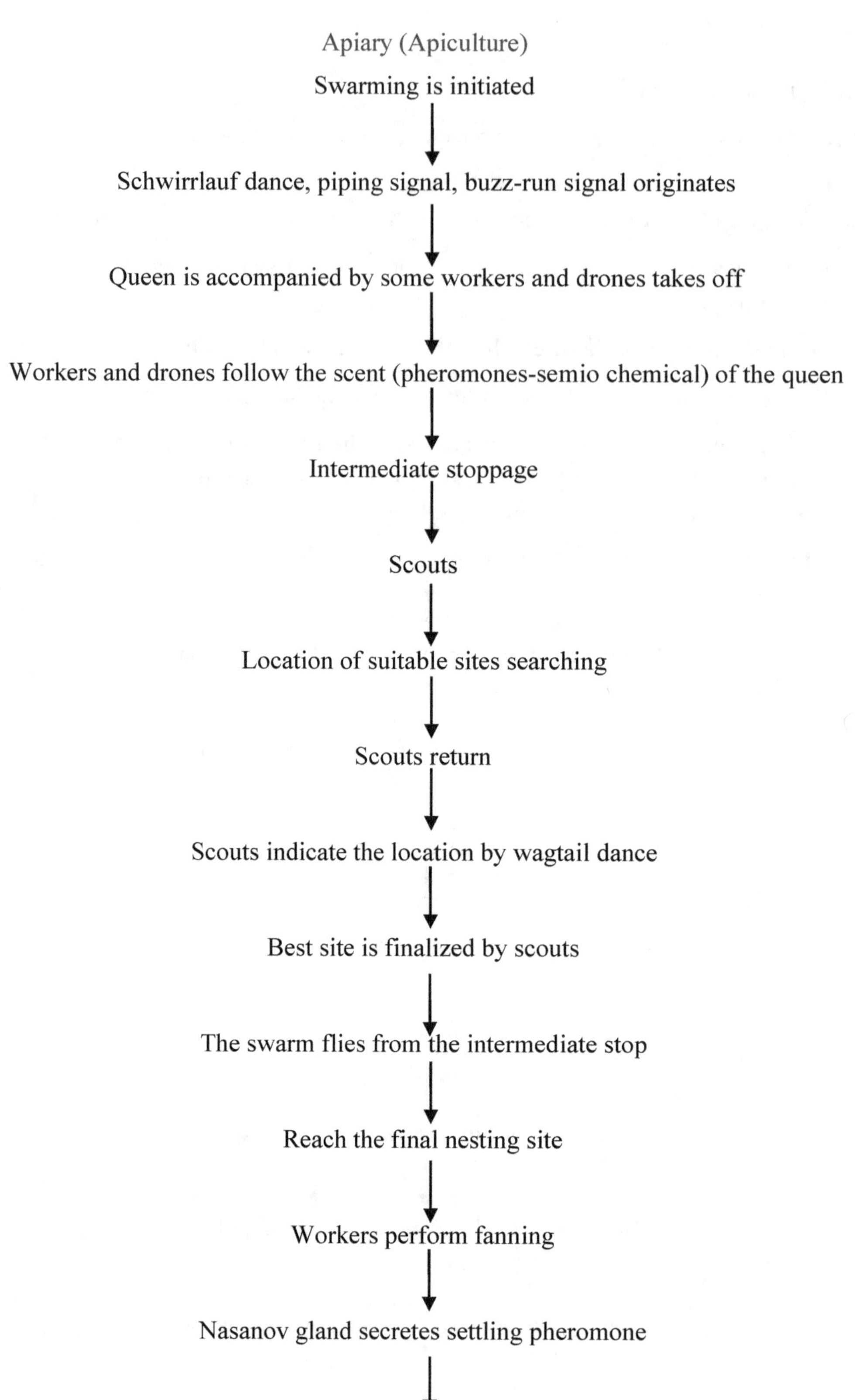

Apiary (Apiculture)
Swarming is initiated
Schwirrlauf dance, piping signal, buzz-run signal originates
Queen is accompanied by some workers and drones takes off
Workers and drones follow the scent (pheromones-semio chemical) of the queen
Intermediate stoppage
Scouts
Location of suitable sites searching
Scouts return
Scouts indicate the location by wagtail dance
Best site is finalized by scouts
The swarm flies from the intermediate stop
Reach the final nesting site
Workers perform fanning
Nasanov gland secretes settling pheromone

She starts egg laying on the 27th day.

If the parent colony is too strong, the daughter queen may also vacant the colony with some workers as an after swarm.

A second daughter queen then takes over the colony in the same way as the first performed.

The swarm which issues from the parent colony is the "primary swarm". The subsequent swarms are known as "after swarms" or casts.

Issue of after swarm depends upon the strength of the colony, the availability of virgin queen etc.

There may be three to four swarmings from a colony depending upon its strength.

If the colony is weak after the issue of the primary swarm, the emerging queen destroys all the brood cells or kills the brood and remain in the parent colony and the activities of the are renewed. Process of swarming has been precisely illustrated in the Fig 8.9.

Incitation of swarming due to overcrowding and less queen pheromone secretion

↓

Poor distribution of queen pheromone secretion

↓

Some worker bees do not receive queen pheromones

↓

They construct queen cups

↓

Queen lays eggs in the queen cups

↓

Queen cell

↓

Egg hatches into larva on the 4th day

↓

Nurse bees feed the larvae with royal jelly

↓

Larvae metamorphoses into pupa on the 8th day

↓

The queen cell is capped

↓

Apiary (Apiculture)

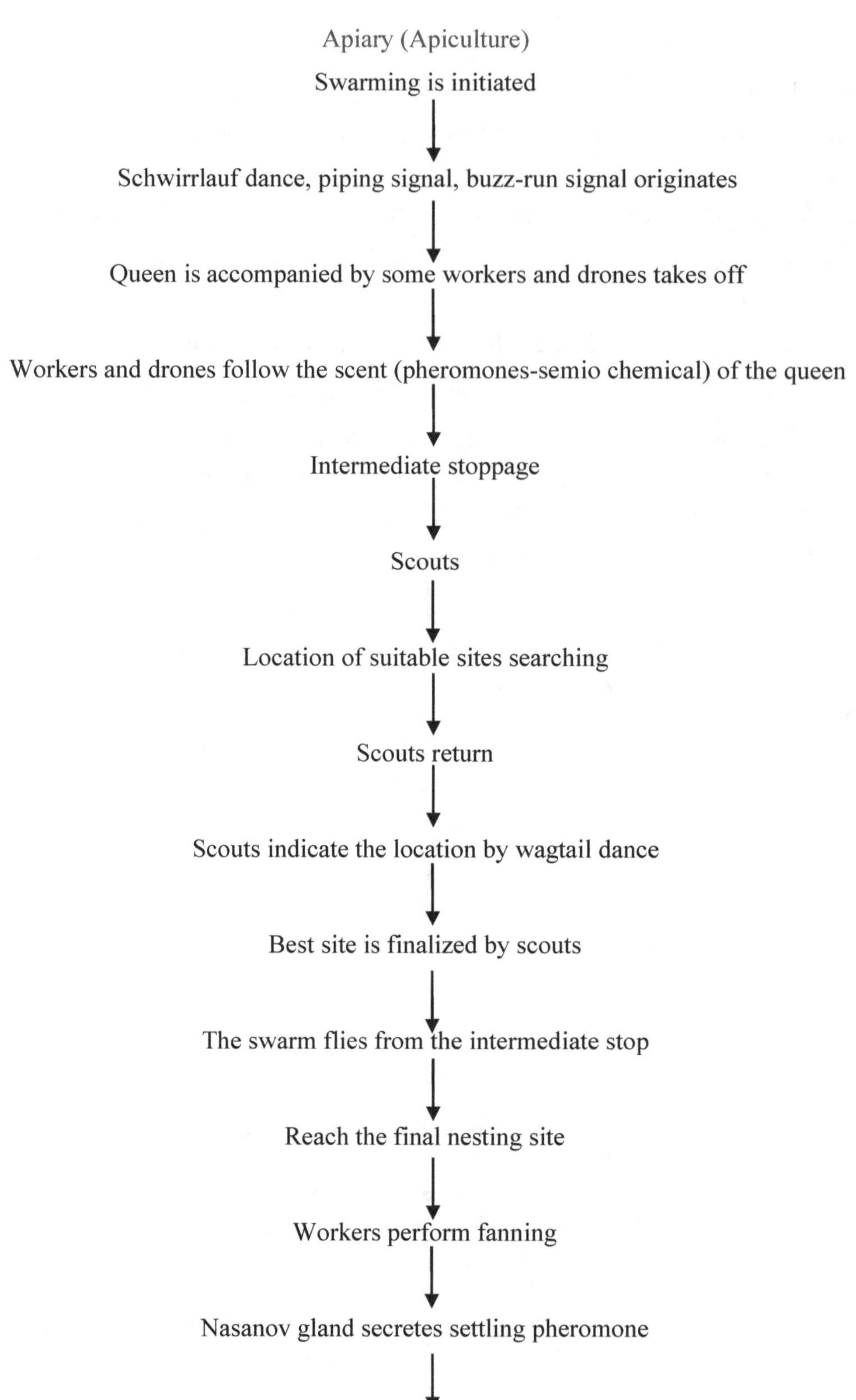

This pheromone communicates the bees "The is our new comb, hurry up and move in".

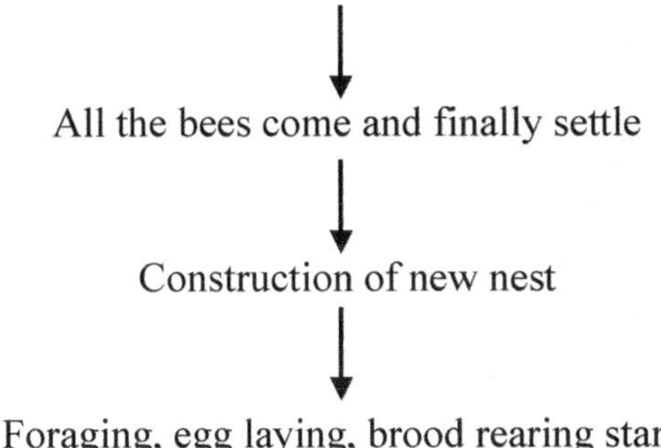

Fig. 8.9 The process of swarming.

Basic types of Swarming

Swarming is of two basic types.
1. Reproductive Swarming and 2. Overcrowd swarming

1. Reproductive Swarming :

It is a method of colony reproduction and for perpetuation of the species. It takes place in the following condition:

- When there are enough bees and enough stores in the hive.
- When the season/weather is favourable for building up the new colony.
- When the conditions are upto the mark to survive the winter without endangering the survival of the colony.

Note- In India, reproductive swarming occurs normally during the seasons with abundant supply of pollen and nectar i.e. (November – March). It does not take place during lean seasons.

This process results in the division of a strong colony into two or more new young colony.

2. Overcrowd swarming

It is a means of controlling the population. It takes place in the following conditions:

- When there are too many bees in the hive.
- When there is not enough space in room.
- When there is not enough stores to continue under the current conditions,
- To improve the colony's chances of survival.

Salient Reasons for Swarming:- Following are the salient reasons for swarming:

1. Inadequate amount of queen pheromone

2. Congestion

3. Becoming of queen weak or old

4. Missing of queen

5. Prolonged periods of bad and adverse weather

6. To replace defective (unproductive) queen

7. Poor ventilation in the comb

8. Scarcity of food or water supply

9. Disease or pest infestation in the hive

Salient Symptoms of Swarming- Following are the salient sysmptoms of swarming:

* An increased rate of egg laying and brood rearing precede swarming.

* Construction of drone cells on the two sides of the comb and drone rearing.

* Construction of swarm cells (queen cells) on the lower edge of the comb. The number of queen cells raised under swarming is more (9) than under supersedure (3) impulse

* Capped queen cells reveals that the swarm has left.

Salient Consequences of Swarming- Following are the salient consequences of swarming**:**

- It is a nuisance to the bee keeper.
- The strength of the colony is reduced.
- The amount of honey and other valuable products that the colony might produç is considerably reduced resulting in serious losses.
- Benefit to the bee keepers is reduced.

Salient Advantages of Swarming- Following are the salient advantages of swarming:

1. It produces new colonies.

2. A new colony can be prepared.

3. Using swarm cells queen rearing can be carried out.

4. Requeening can be done.

5. New colony is established.

Swarm Control

Swarming is a loss to the beekeeper and a nuisance to the beekeeper and public. Hence, it has to be controlled.

The prevention or checking up of swarming is called swarm control.

Salient Factors which aid to control Swarming- Following are the salient factors which assist in control of swarming:

1. Bee keepers should select such species which has less tendency for swarming. *Apis mellifera* has less tendency for swarming than *Apis indica.*

2. An young queen is used to head the colony.

3. Brood combs of the brood nest must be removed and replaced by empty combs to assist the queen laying the eggs.

4. Provide ample space in the brood and super chambers.

5. Provide shade to reduce the heat in the hive.

6. Provide ventilation by placing a stick between the lid and the frames.

7. Provide water.

8. Reduce drone rearing by providing worker cell foundation.

9. Destroy queen cells built on the sides or base, because it is an indication that the bees are being ready for swarming.

10. Uncapped and capped queen cells should be cut out and removed. It will postpone the swarming for a week or two.

11. Divide the colony into two if the bees continue to build queen cells.

12. The old queen is destroyed and a new queen is introduced.

13. Use of a queen gate infront of the hive.

14. Congestion stimulates swarming. Hence congestion in the hive can be controlled in the following ways:

* Primary swarm is allowed to take place, but trapped in a swarm trap and hived in a separate colony. The after swarms are prevented by destroying the remaining queen.

* One or two brood combs in the strong colonies which are inclined to issue swarms are removed and given to weak colonies.

* A brood comb with a reigning queen and a few workers are taken out and put in a separate hive and thus the colony is divided.

* Interchange of position between a strong and a weak colony.

Salient Methods of Swarm Control
Swarming may be controlled by the following methods:

1. The demaree method

2. Clipping off the wings

3. Artificial swarming

The Demaree Method
Demaree method is for preventing swarming by separating the queen from the stand.

This method was devised by George Demaree in 1884.

It is done by manipulating the hive and process is called Demareeing.

It requires a new brood chamber with brood comb frames. It is placed between the stand and the brood chamber of the hive.

This method is carried out in following steps:

1. Removal of the hive components one by one from the stand and placing it on the ground in the cited order as below:

Roof

Crown board

Super chamber

Queen excluder

Brood chamber

2. Placement of a new brood box on the stand.

The entire steps of demaree method is precisely shown in the Fig 8.10

Removal of hive from the stand

Components of the hive are separated and placed on the ground in the following order:

Roof → Crown board→ Super chamber → Queen excluder →Brood chamber

↓

Identify the queen in the brood chamber

↓

Removal of the queen from the brood comb

↓

Placing a new brood chamber on the stand

↓

Place the queen containing brood comb in side the new brood chamber

↓

Adding of 2 more brood frames containing eggs and larvae

↓

Filling the empty space with comb frames

↓

Placing the queen excluder above the new brood chamber

↓

Placing the parent super chamber over the queen excluder

↓

Placing the parent brood chamber above the super chamber

↓

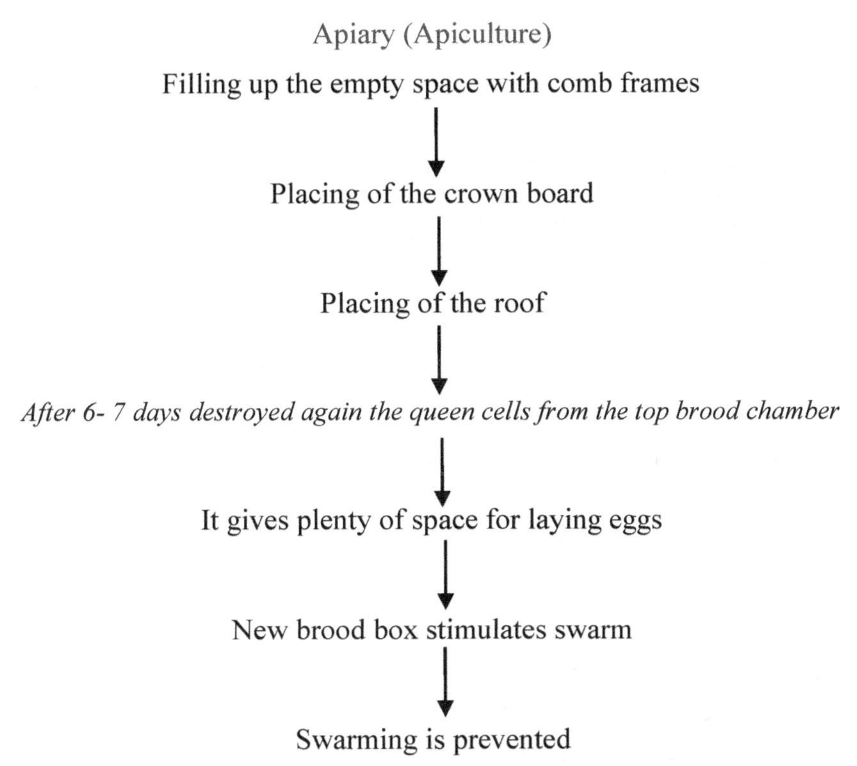

Fig. 8.10 Demaree method of preventing Swarm.

3. Isolate the queen from the original brood box.

4. Place her with two frames of brood in the centre of the new brood' box. Fill up the left space of the new brood chamber with clean empty combs.

5. Place the queen excluder on this new brood box and put the super chamber above the excluder.

6. Put the original brood box with the rest brood on top. Fill up the gap with 2 frames of crown comb.

7. Fit crown board and roof on the beehive.

8. In about 30 minutes most of the nurse bees migrate to the brood chamber at the top and the other foraging bees stay with the queen below the queen excluder, as if they had swarmed.

9. After 7-8 days, examine the top brood box. These queen cells are destroyed.

The beehive is stacked as follows:

Roof
Crown board
Original-brood box having with young brood
Super chamber
New Brood box with queen, 2 frames sealed brood and rest empty combs

10. In 20-21 days, all the brood will be hatched out of the combs above the excluder. New bees will begin to hatch in the queen's chamber below the excluder. Therefore a continued succession of young bees is sustained with no set back to the colony or loss of honey production.

11. The method of separating most brood from the queen is repeated after days with queen cells destroyed in the top box.

12. By performing this, the queen may have plenty of space to lay eggs and expend her brood nest.

13. This procedure reduces congestion in the lower brood box because most of the nurse bees are observed in upstairs.

14. The foraging bees may get a great deal of room to store honey in the middle hive bodies.

The salient features of this system are the following:

- The colony has all of its brood and the queen.
- The queen is separated from most of her brood.
- The queen has a new nest below the excluder.
- Major portion of the bees is in the top where most of the brood has been moved, to.
- A minor portion of the bees with a small portion of brood and queen are in the bottom brood box.

2. Clipping of the wings

Clipping the wings of queen checks her from flying away with the swarm. If the old queen has failed to leave with the swarm, the bees will return to their hive and wait until virgin has emerged. Therefore, clipping the wings only stops swarming temporarily.

3. Artificial swarming

Artificial swarming is a swarm created by a beekeeper from a large colony which is about to swarm.

In fact, a swarm is a mass of bees having a queen, workers and drones.

The swarm constituted by the bee keeper is called artificial swarm and the process of creating artificial swarm is called artificial swarming.

Advantages of artificial swarming: Artificial swarming has the following advantages:

* It prevents natural swarming.

* It assists to produce new colonies.

* It assists in requeening.

The creation of artificial swarm needs a new small hive having the following assets:

* Hive stand

* Bottom board

* Brood chamber with comb frames

* Crown board

* Roof

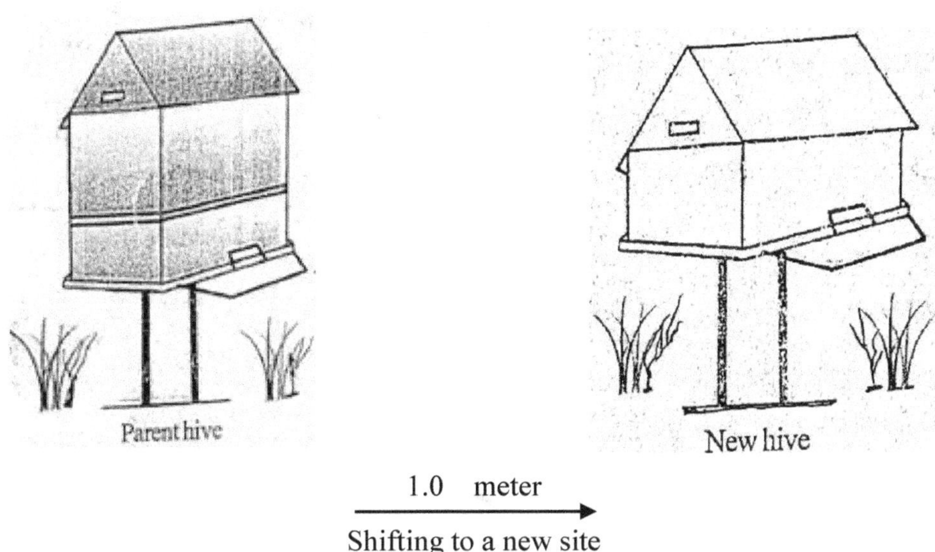

Fig. 8.11: Creation of artificial swarm.

Creation of artificial swarms

The creation of artificial swarms involves the following steps:

1. Examine the parent colony for queen cells. If the colony has queen cells, it is ready for swarming.

2. Shift the parent hive from its original place to a new site at least one meter away.

3. Set the new hive in the original site of the parent colony.

4. Observe the location of the queen in the parent colony.

5. Remove the queen from the brood frame.

6. Remove 3 comb frames from the new hive.

7. Shift the queen with the comb frame in the brood chamber of the new hive.

8. Remove 2 more brood frames from the parent colony and shift them in the new hive on side of the queen frame.

9. Now the new hive is known as the artificial swarm. The queen is now in the artificial swarm. The flying bees back to the new hive as it is on the original site. This mimics what happens when the bees swarm naturally where the queen leaves the original hive along with the flying bees.

10. The parent hive is now without queen. But a number of queen cells are present. The non flying house bees pay attention after the brood just as when a natural swarm occurs. New queens will emerge soon.

11. Fill the gap in the parent brood chamber by inserting 3 comb frames.

12. Now leave both hives as such.

13. Both colonies may be reared as separate colonies or united into single stock.

Important Question:

1. Give an account of social behavior of honey bees.

2. Define communication. Describe pattern of communication in honey bees with special reference to dance language.

3. Define swarming. Give its types and control of swarming.

4. Write short notes on the following:

   A. Queen
   B. Drones
   C. Workers
   D. Social organization in bee colony
   E. Swarming behavior of honey bee
   F. Waggle dance

Suggested Readings

1. Applied Zoology- By N. Arumugam; T. Morugam; J. Johnson Rajeshwar; A. Ram Prabhu: Saras Publications.

2. Fundamentals of Animal Behaviour- By J.P. Shukla; Atlantic publication; New Delhi.

3. Economic Zoology- By G.S. Shukla and V.B. Upadhyay; Rastogi Publications.

4. A Text Book of Apiculture- By Hemraj; S. Vinash and Co.

5. Text Book on Bee Keeping- By Ataur Rahman; Kalyani Publication.

6. Understating Apiculture- By Ashok Kumar; Discovery Publishing Pvt. Ltd.

7. A Text Book of Apiculture- By B.S. Pagar; Sahitya Sagar.

8. Apiculture in India- By Ataur Rahman; ICAR, New Delhi.

9. Fundamental of Bee Keeping- By T.V. Sathe; Daya Publishing House.

10. Bee Keeping For Beginners- By Erin Morrow; Mihailis Konoplous.

11. Theare are Queen Cells in my hive- what should I do?- By Shaw Wally; Northern Bee Books.

12. New Bee Keeping in a Long Deep Hive- By Darington Robin; Northern Bee Books.